Artificial Intelligence: A Very Short Introduction

VERY SHORT INTRODUCTIONS are for anyone wanting a stimulating and accessible way into a new subject. They are written by experts, and have been translated into more than 45 different languages.

The series began in 1995, and now covers a wide variety of topics in every discipline. The VSI library currently contains over 550 volumes—a Very Short Introduction to everything from Psychology and Philosophy of Science to American History and Relativity—and continues to grow in every subject area.

Very Short Introductions available now:

ACCOUNTING Christopher Nobes
ADOLESCENCE Peter K. Smith
ADVERTISING Winston Fletcher
AFRICAN AMERICAN RELIGION Eddie S. Glaude Jr
AFRICAN HISTORY John Parker and Richard Rathbone
AFRICAN RELIGIONS Jacob K. Olupona
AGEING Nancy A. Pachana
AGNOSTICISM Robin Le Poidevin
AGRICULTURE Paul Brassley and Richard Soffe
ALEXANDER THE GREAT Hugh Bowden
ALGEBRA Peter M. Higgins
AMERICAN HISTORY Paul S. Boyer
AMERICAN IMMIGRATION David A. Gerber
AMERICAN LEGAL HISTORY G. Edward White
AMERICAN POLITICAL HISTORY Donald Critchlow
AMERICAN POLITICAL PARTIES AND ELECTIONS L. Sandy Maisel
AMERICAN POLITICS Richard M. Valelly
THE AMERICAN PRESIDENCY Charles O. Jones
THE AMERICAN REVOLUTION Robert J. Allison
AMERICAN SLAVERY Heather Andrea Williams
THE AMERICAN WEST Stephen Aron
AMERICAN WOMEN'S HISTORY Susan Ware
ANAESTHESIA Aidan O'Donnell
ANALYTIC PHILOSOPHY Michael Beaney
ANARCHISM Colin Ward
ANCIENT ASSYRIA Karen Radner
ANCIENT EGYPT Ian Shaw
ANCIENT EGYPTIAN ART AND ARCHITECTURE Christina Riggs
ANCIENT GREECE Paul Cartledge
THE ANCIENT NEAR EAST Amanda H. Podany
ANCIENT PHILOSOPHY Julia Annas
ANCIENT WARFARE Harry Sidebottom
ANGELS David Albert Jones
ANGLICANISM Mark Chapman
THE ANGLO-SAXON AGE John Blair
ANIMAL BEHAVIOUR Tristram D. Wyatt
THE ANIMAL KINGDOM Peter Holland
ANIMAL RIGHTS David DeGrazia
THE ANTARCTIC Klaus Dodds
ANTHROPOCENE Erle C. Ellis
ANTISEMITISM Steven Beller
ANXIETY Daniel Freeman and Jason Freeman
APPLIED MATHEMATICS Alain Goriely
THE APOCRYPHAL GOSPELS Paul Foster
ARCHAEOLOGY Paul Bahn
ARCHITECTURE Andrew Ballantyne
ARISTOCRACY William Doyle
ARISTOTLE Jonathan Barnes

ART HISTORY Dana Arnold
ART THEORY Cynthia Freeland
ARTIFICIAL INTELLIGENCE Margaret A. Boden
ASIAN AMERICAN HISTORY Madeline Y. Hsu
ASTROBIOLOGY David C. Catling
ASTROPHYSICS James Binney
ATHEISM Julian Baggini
THE ATMOSPHERE Paul I. Palmer
AUGUSTINE Henry Chadwick
AUSTRALIA Kenneth Morgan
AUTISM Uta Frith
AUTOBIOGRAPHY Laura Marcus
THE AVANT GARDE David Cottington
THE AZTECS Davíd Carrasco
BABYLONIA Trevor Bryce
BACTERIA Sebastian G. B. Amyes
BANKING John Goddard and John O. S. Wilson
BARTHES Jonathan Culler
THE BEATS David Sterritt
BEAUTY Roger Scruton
BEHAVIOURAL ECONOMICS Michelle Baddeley
BESTSELLERS John Sutherland
THE BIBLE John Riches
BIBLICAL ARCHAEOLOGY Eric H. Cline
BIG DATA Dawn E. Holmes
BIOGRAPHY Hermione Lee
BLACK HOLES Katherine Blundell
BLOOD Chris Cooper
THE BLUES Elijah Wald
THE BODY Chris Shilling
THE BOOK OF MORMON Terryl Givens
BORDERS Alexander C. Diener and Joshua Hagen
THE BRAIN Michael O'Shea
BRANDING Robert Jones
THE BRICS Andrew F. Cooper
THE BRITISH CONSTITUTION Martin Loughlin
THE BRITISH EMPIRE Ashley Jackson
BRITISH POLITICS Anthony Wright
BUDDHA Michael Carrithers
BUDDHISM Damien Keown
BUDDHIST ETHICS Damien Keown
BYZANTIUM Peter Sarris
CALVINISM Jon Balserak
CANCER Nicholas James
CAPITALISM James Fulcher
CATHOLICISM Gerald O'Collins
CAUSATION Stephen Mumford and Rani Lill Anjum
THE CELL Terence Allen and Graham Cowling
THE CELTS Barry Cunliffe
CHAOS Leonard Smith
CHEMISTRY Peter Atkins
CHILD PSYCHOLOGY Usha Goswami
CHILDREN'S LITERATURE Kimberley Reynolds
CHINESE LITERATURE Sabina Knight
CHOICE THEORY Michael Allingham
CHRISTIAN ART Beth Williamson
CHRISTIAN ETHICS D. Stephen Long
CHRISTIANITY Linda Woodhead
CIRCADIAN RHYTHMS Russell Foster and Leon Kreitzman
CITIZENSHIP Richard Bellamy
CIVIL ENGINEERING David Muir Wood
CLASSICAL LITERATURE William Allan
CLASSICAL MYTHOLOGY Helen Morales
CLASSICS Mary Beard and John Henderson
CLAUSEWITZ Michael Howard
CLIMATE Mark Maslin
CLIMATE CHANGE Mark Maslin
CLINICAL PSYCHOLOGY Susan Llewelyn and Katie Aafjes-van Doorn
COGNITIVE NEUROSCIENCE Richard Passingham
THE COLD WAR Robert McMahon
COLONIAL AMERICA Alan Taylor
COLONIAL LATIN AMERICAN LITERATURE Rolena Adorno
COMBINATORICS Robin Wilson
COMEDY Matthew Bevis
COMMUNISM Leslie Holmes
COMPARATIVE LITERATURE Ben Hutchinson
COMPLEXITY John H. Holland
THE COMPUTER Darrel Ince
COMPUTER SCIENCE Subrata Dasgupta
CONFUCIANISM Daniel K. Gardner

THE CONQUISTADORS Matthew Restall and Felipe Fernández-Armesto
CONSCIENCE Paul Strohm
CONSCIOUSNESS Susan Blackmore
CONTEMPORARY ART Julian Stallabrass
CONTEMPORARY FICTION Robert Eaglestone
CONTINENTAL PHILOSOPHY Simon Critchley
COPERNICUS Owen Gingerich
CORAL REEFS Charles Sheppard
CORPORATE SOCIAL RESPONSIBILITY Jeremy Moon
CORRUPTION Leslie Holmes
COSMOLOGY Peter Coles
CRIME FICTION Richard Bradford
CRIMINAL JUSTICE Julian V. Roberts
CRIMINOLOGY Tim Newburn
CRITICAL THEORY Stephen Eric Bronner
THE CRUSADES Christopher Tyerman
CRYPTOGRAPHY Fred Piper and Sean Murphy
CRYSTALLOGRAPHY A. M. Glazer
THE CULTURAL REVOLUTION Richard Curt Kraus
DADA AND SURREALISM David Hopkins
DANTE Peter Hainsworth and David Robey
DARWIN Jonathan Howard
THE DEAD SEA SCROLLS Timothy H. Lim
DECOLONIZATION Dane Kennedy
DEMOCRACY Bernard Crick
DEMOGRAPHY Sarah Harper
DEPRESSION Jan Scott and Mary Jane Tacchi
DERRIDA Simon Glendinning
DESCARTES Tom Sorell
DESERTS Nick Middleton
DESIGN John Heskett
DEVELOPMENT Ian Goldin
DEVELOPMENTAL BIOLOGY Lewis Wolpert
THE DEVIL Darren Oldridge
DIASPORA Kevin Kenny
DICTIONARIES Lynda Mugglestone
DINOSAURS David Norman
DIPLOMACY Joseph M. Siracusa
DOCUMENTARY FILM Patricia Aufderheide
DREAMING J. Allan Hobson
DRUGS Les Iversen
DRUIDS Barry Cunliffe
EARLY MUSIC Thomas Forrest Kelly
THE EARTH Martin Redfern
EARTH SYSTEM SCIENCE Tim Lenton
ECONOMICS Partha Dasgupta
EDUCATION Gary Thomas
EGYPTIAN MYTH Geraldine Pinch
EIGHTEENTH-CENTURY BRITAIN Paul Langford
THE ELEMENTS Philip Ball
EMOTION Dylan Evans
EMPIRE Stephen Howe
ENGELS Terrell Carver
ENGINEERING David Blockley
THE ENGLISH LANGUAGE Simon Horobin
ENGLISH LITERATURE Jonathan Bate
THE ENLIGHTENMENT John Robertson
ENTREPRENEURSHIP Paul Westhead and Mike Wright
ENVIRONMENTAL ECONOMICS Stephen Smith
ENVIRONMENTAL LAW Elizabeth Fisher
ENVIRONMENTAL POLITICS Andrew Dobson
EPICUREANISM Catherine Wilson
EPIDEMIOLOGY Rodolfo Saracci
ETHICS Simon Blackburn
ETHNOMUSICOLOGY Timothy Rice
THE ETRUSCANS Christopher Smith
EUGENICS Philippa Levine
THE EUROPEAN UNION Simon Usherwood and John Pinder
EUROPEAN UNION LAW Anthony Arnull
EVOLUTION Brian and Deborah Charlesworth
EXISTENTIALISM Thomas Flynn
EXPLORATION Stewart A. Weaver
THE EYE Michael Land
FAIRY TALE Marina Warner
FAMILY LAW Jonathan Herring
FASCISM Kevin Passmore

FASHION Rebecca Arnold
FEMINISM Margaret Walters
FILM Michael Wood
FILM MUSIC Kathryn Kalinak
THE FIRST WORLD WAR Michael Howard
FOLK MUSIC Mark Slobin
FOOD John Krebs
FORENSIC PSYCHOLOGY David Canter
FORENSIC SCIENCE Jim Fraser
FORESTS Jaboury Ghazoul
FOSSILS Keith Thomson
FOUCAULT Gary Gutting
THE FOUNDING FATHERS R. B. Bernstein
FRACTALS Kenneth Falconer
FREE SPEECH Nigel Warburton
FREE WILL Thomas Pink
FREEMASONRY Andreas Önnerfors
FRENCH LITERATURE John D. Lyons
THE FRENCH REVOLUTION William Doyle
FREUD Anthony Storr
FUNDAMENTALISM Malise Ruthven
FUNGI Nicholas P. Money
THE FUTURE Jennifer M. Gidley
GALAXIES John Gribbin
GALILEO Stillman Drake
GAME THEORY Ken Binmore
GANDHI Bhikhu Parekh
GENES Jonathan Slack
GENIUS Andrew Robinson
GENOMICS John Archibald
GEOGRAPHY John Matthews and David Herbert
GEOLOGY Jan Zalasiewicz
GEOPHYSICS William Lowrie
GEOPOLITICS Klaus Dodds
GERMAN LITERATURE Nicholas Boyle
GERMAN PHILOSOPHY Andrew Bowie
GLOBAL CATASTROPHES Bill McGuire
GLOBAL ECONOMIC HISTORY Robert C. Allen
GLOBALIZATION Manfred Steger
GOD John Bowker
GOETHE Ritchie Robertson
THE GOTHIC Nick Groom
GOVERNANCE Mark Bevir
GRAVITY Timothy Clifton
THE GREAT DEPRESSION AND THE NEW DEAL Eric Rauchway
HABERMAS James Gordon Finlayson
THE HABSBURG EMPIRE Martyn Rady
HAPPINESS Daniel M. Haybron
THE HARLEM RENAISSANCE Cheryl A. Wall
THE HEBREW BIBLE AS LITERATURE Tod Linafelt
HEGEL Peter Singer
HEIDEGGER Michael Inwood
THE HELLENISTIC AGE Peter Thonemann
HEREDITY John Waller
HERMENEUTICS Jens Zimmermann
HERODOTUS Jennifer T. Roberts
HIEROGLYPHS Penelope Wilson
HINDUISM Kim Knott
HISTORY John H. Arnold
THE HISTORY OF ASTRONOMY Michael Hoskin
THE HISTORY OF CHEMISTRY William H. Brock
THE HISTORY OF CINEMA Geoffrey Nowell-Smith
THE HISTORY OF LIFE Michael Benton
THE HISTORY OF MATHEMATICS Jacqueline Stedall
THE HISTORY OF MEDICINE William Bynum
THE HISTORY OF PHYSICS J. L. Heilbron
THE HISTORY OF TIME Leofranc Holford-Strevens
HIV AND AIDS Alan Whiteside
HOBBES Richard Tuck
HOLLYWOOD Peter Decherney
THE HOLY ROMAN EMPIRE Joachim Whaley
HOME Michael Allen Fox
HORMONES Martin Luck
HUMAN ANATOMY Leslie Klenerman
HUMAN EVOLUTION Bernard Wood
HUMAN RIGHTS Andrew Clapham
HUMANISM Stephen Law
HUME A. J. Ayer
HUMOUR Noël Carroll
THE ICE AGE Jamie Woodward

IDEOLOGY Michael Freeden
THE IMMUNE SYSTEM Paul Klenerman
INDIAN CINEMA Ashish Rajadhyaksha
INDIAN PHILOSOPHY Sue Hamilton
THE INDUSTRIAL REVOLUTION Robert C. Allen
INFECTIOUS DISEASE Marta L. Wayne and Benjamin M. Bolker
INFINITY Ian Stewart
INFORMATION Luciano Floridi
INNOVATION Mark Dodgson and David Gann
INTELLIGENCE Ian J. Deary
INTELLECTUAL PROPERTY Siva Vaidhyanathan
INTERNATIONAL LAW Vaughan Lowe
INTERNATIONAL MIGRATION Khalid Koser
INTERNATIONAL RELATIONS Paul Wilkinson
INTERNATIONAL SECURITY Christopher S. Browning
IRAN Ali M. Ansari
ISLAM Malise Ruthven
ISLAMIC HISTORY Adam Silverstein
ISOTOPES Rob Ellam
ITALIAN LITERATURE Peter Hainsworth and David Robey
JESUS Richard Bauckham
JEWISH HISTORY David N. Myers
JOURNALISM Ian Hargreaves
JUDAISM Norman Solomon
JUNG Anthony Stevens
KABBALAH Joseph Dan
KAFKA Ritchie Robertson
KANT Roger Scruton
KEYNES Robert Skidelsky
KIERKEGAARD Patrick Gardiner
KNOWLEDGE Jennifer Nagel
THE KORAN Michael Cook
LAKES Warwick F. Vincent
LANDSCAPE ARCHITECTURE Ian H. Thompson
LANDSCAPES AND GEOMORPHOLOGY Andrew Goudie and Heather Viles
LANGUAGES Stephen R. Anderson
LATE ANTIQUITY Gillian Clark
LAW Raymond Wacks
THE LAWS OF THERMODYNAMICS Peter Atkins
LEADERSHIP Keith Grint
LEARNING Mark Haselgrove
LEIBNIZ Maria Rosa Antognazza
LIBERALISM Michael Freeden
LIGHT Ian Walmsley
LINCOLN Allen C. Guelzo
LINGUISTICS Peter Matthews
LITERARY THEORY Jonathan Culler
LOCKE John Dunn
LOGIC Graham Priest
LOVE Ronald de Sousa
MACHIAVELLI Quentin Skinner
MADNESS Andrew Scull
MAGIC Owen Davies
MAGNA CARTA Nicholas Vincent
MAGNETISM Stephen Blundell
MALTHUS Donald Winch
MAMMALS T. S. Kemp
MANAGEMENT John Hendry
MAO Delia Davin
MARINE BIOLOGY Philip V. Mladenov
THE MARQUIS DE SADE John Phillips
MARTIN LUTHER Scott H. Hendrix
MARTYRDOM Jolyon Mitchell
MARX Peter Singer
MATERIALS Christopher Hall
MATHEMATICS Timothy Gowers
THE MEANING OF LIFE Terry Eagleton
MEASUREMENT David Hand
MEDICAL ETHICS Tony Hope
MEDICAL LAW Charles Foster
MEDIEVAL BRITAIN John Gillingham and Ralph A. Griffiths
MEDIEVAL LITERATURE Elaine Treharne
MEDIEVAL PHILOSOPHY John Marenbon
MEMORY Jonathan K. Foster
METAPHYSICS Stephen Mumford
THE MEXICAN REVOLUTION Alan Knight
MICHAEL FARADAY Frank A. J. L. James
MICROBIOLOGY Nicholas P. Money
MICROECONOMICS Avinash Dixit
MICROSCOPY Terence Allen
THE MIDDLE AGES Miri Rubin

MILITARY JUSTICE Eugene R. Fidell
MILITARY STRATEGY Antulio J. Echevarria II
MINERALS David Vaughan
MIRACLES Yujin Nagasawa
MODERN ART David Cottington
MODERN CHINA Rana Mitter
MODERN DRAMA Kirsten E. Shepherd-Barr
MODERN FRANCE Vanessa R. Schwartz
MODERN INDIA Craig Jeffrey
MODERN IRELAND Senia Pašeta
MODERN ITALY Anna Cento Bull
MODERN JAPAN Christopher Goto-Jones
MODERN LATIN AMERICAN LITERATURE Roberto González Echevarría
MODERN WAR Richard English
MODERNISM Christopher Butler
MOLECULAR BIOLOGY Aysha Divan and Janice A. Royds
MOLECULES Philip Ball
MONASTICISM Stephen J. Davis
THE MONGOLS Morris Rossabi
MOONS David A. Rothery
MORMONISM Richard Lyman Bushman
MOUNTAINS Martin F. Price
MUHAMMAD Jonathan A. C. Brown
MULTICULTURALISM Ali Rattansi
MULTILINGUALISM John C. Maher
MUSIC Nicholas Cook
MYTH Robert A. Segal
THE NAPOLEONIC WARS Mike Rapport
NATIONALISM Steven Grosby
NATIVE AMERICAN LITERATURE Sean Teuton
NAVIGATION Jim Bennett
NELSON MANDELA Elleke Boehmer
NEOLIBERALISM Manfred Steger and Ravi Roy
NETWORKS Guido Caldarelli and Michele Catanzaro
THE NEW TESTAMENT Luke Timothy Johnson
THE NEW TESTAMENT AS LITERATURE Kyle Keefer
NEWTON Robert Iliffe
NIETZSCHE Michael Tanner
NINETEENTH-CENTURY BRITAIN Christopher Harvie and H. C. G. Matthew
THE NORMAN CONQUEST George Garnett
NORTH AMERICAN INDIANS Theda Perdue and Michael D. Green
NORTHERN IRELAND Marc Mulholland
NOTHING Frank Close
NUCLEAR PHYSICS Frank Close
NUCLEAR POWER Maxwell Irvine
NUCLEAR WEAPONS Joseph M. Siracusa
NUMBERS Peter M. Higgins
NUTRITION David A. Bender
OBJECTIVITY Stephen Gaukroger
OCEANS Dorrik Stow
THE OLD TESTAMENT Michael D. Coogan
THE ORCHESTRA D. Kern Holoman
ORGANIC CHEMISTRY Graham Patrick
ORGANIZED CRIME Georgios A. Antonopoulos and Georgios Papanicolaou
ORGANIZATIONS Mary Jo Hatch
PAGANISM Owen Davies
PAIN Rob Boddice
THE PALESTINIAN-ISRAELI CONFLICT Martin Bunton
PANDEMICS Christian W. McMillen
PARTICLE PHYSICS Frank Close
PAUL E. P. Sanders
PEACE Oliver P. Richmond
PENTECOSTALISM William K. Kay
PERCEPTION Brian Rogers
THE PERIODIC TABLE Eric R. Scerri
PHILOSOPHY Edward Craig
PHILOSOPHY IN THE ISLAMIC WORLD Peter Adamson
PHILOSOPHY OF LAW Raymond Wacks
PHILOSOPHY OF SCIENCE Samir Okasha
PHILOSOPHY OF RELIGION Tim Bayne
PHOTOGRAPHY Steve Edwards
PHYSICAL CHEMISTRY Peter Atkins
PILGRIMAGE Ian Reader

PLAGUE Paul Slack
PLANETS David A. Rothery
PLANTS Timothy Walker
PLATE TECTONICS Peter Molnar
PLATO Julia Annas
POLITICAL PHILOSOPHY David Miller
POLITICS Kenneth Minogue
POPULISM Cas Mudde and Cristóbal Rovira Kaltwasser
POSTCOLONIALISM Robert Young
POSTMODERNISM Christopher Butler
POSTSTRUCTURALISM Catherine Belsey
POVERTY Philip N. Jefferson
PREHISTORY Chris Gosden
PRESOCRATIC PHILOSOPHY Catherine Osborne
PRIVACY Raymond Wacks
PROBABILITY John Haigh
PROGRESSIVISM Walter Nugent
PROJECTS Andrew Davies
PROTESTANTISM Mark A. Noll
PSYCHIATRY Tom Burns
PSYCHOANALYSIS Daniel Pick
PSYCHOLOGY Gillian Butler and Freda McManus
PSYCHOTHERAPY Tom Burns and Eva Burns-Lundgren
PUBLIC ADMINISTRATION Stella Z. Theodoulou and Ravi K. Roy
PUBLIC HEALTH Virginia Berridge
PURITANISM Francis J. Bremer
THE QUAKERS Pink Dandelion
QUANTUM THEORY John Polkinghorne
RACISM Ali Rattansi
RADIOACTIVITY Claudio Tuniz
RASTAFARI Ennis B. Edmonds
THE REAGAN REVOLUTION Gil Troy
REALITY Jan Westerhoff
THE REFORMATION Peter Marshall
RELATIVITY Russell Stannard
RELIGION IN AMERICA Timothy Beal
THE RENAISSANCE Jerry Brotton
RENAISSANCE ART Geraldine A. Johnson
REVOLUTIONS Jack A. Goldstone
RHETORIC Richard Toye
RISK Baruch Fischhoff and John Kadvany
RITUAL Barry Stephenson
RIVERS Nick Middleton
ROBOTICS Alan Winfield
ROCKS Jan Zalasiewicz
ROMAN BRITAIN Peter Salway
THE ROMAN EMPIRE Christopher Kelly
THE ROMAN REPUBLIC David M. Gwynn
ROMANTICISM Michael Ferber
ROUSSEAU Robert Wokler
RUSSELL A. C. Grayling
RUSSIAN HISTORY Geoffrey Hosking
RUSSIAN LITERATURE Catriona Kelly
THE RUSSIAN REVOLUTION S. A. Smith
SAVANNAS Peter A. Furley
SCHIZOPHRENIA Chris Frith and Eve Johnstone
SCHOPENHAUER Christopher Janaway
SCIENCE AND RELIGION Thomas Dixon
SCIENCE FICTION David Seed
THE SCIENTIFIC REVOLUTION Lawrence M. Principe
SCOTLAND Rab Houston
SEXUAL SELECTION Marlene Zuk and Leigh W. Simmons
SEXUALITY Véronique Mottier
SHAKESPEARE'S COMEDIES Bart van Es
SHAKESPEARE'S SONNETS AND POEMS Jonathan F. S. Post
SHAKESPEARE'S TRAGEDIES Stanley Wells
SIKHISM Eleanor Nesbitt
THE SILK ROAD James A. Millward
SLANG Jonathon Green
SLEEP Steven W. Lockley and Russell G. Foster
SOCIAL AND CULTURAL ANTHROPOLOGY John Monaghan and Peter Just
SOCIAL PSYCHOLOGY Richard J. Crisp
SOCIAL WORK Sally Holland and Jonathan Scourfield
SOCIALISM Michael Newman
SOCIOLINGUISTICS John Edwards
SOCIOLOGY Steve Bruce
SOCRATES C. C. W. Taylor
SOUND Mike Goldsmith

THE SOVIET UNION Stephen Lovell
THE SPANISH CIVIL WAR
Helen Graham
SPANISH LITERATURE Jo Labanyi
SPINOZA Roger Scruton
SPIRITUALITY Philip Sheldrake
SPORT Mike Cronin
STARS Andrew King
STATISTICS David J. Hand
STEM CELLS Jonathan Slack
STOICISM Brad Inwood
STRUCTURAL ENGINEERING
David Blockley
STUART BRITAIN John Morrill
SUPERCONDUCTIVITY
Stephen Blundell
SYMMETRY Ian Stewart
SYNTHETIC BIOLOGY Jamie A. Davies
TAXATION Stephen Smith
TEETH Peter S. Ungar
TELESCOPES Geoff Cottrell
TERRORISM Charles Townshend
THEATRE Marvin Carlson
THEOLOGY David F. Ford
THINKING AND REASONING
Jonathan St B. T. Evans
THOMAS AQUINAS Fergus Kerr
THOUGHT Tim Bayne
TIBETAN BUDDHISM
Matthew T. Kapstein
TOCQUEVILLE Harvey C. Mansfield
TRAGEDY Adrian Poole
TRANSLATION Matthew Reynolds
THE TROJAN WAR Eric H. Cline
TRUST Katherine Hawley
THE TUDORS John Guy
TWENTIETH-CENTURY BRITAIN
Kenneth O. Morgan
THE UNITED NATIONS
Jussi M. Hanhimäki
THE U.S. CONGRESS
Donald A. Ritchie
THE U.S. CONSTITUTION
David J. Bodenhamer
THE U.S. SUPREME COURT
Linda Greenhouse
UTILITARIANISM
Katarzyna de Lazari-Radek
and Peter Singer
UNIVERSITIES AND COLLEGES
David Palfreyman and Paul Temple
UTOPIANISM Lyman Tower Sargent
VETERINARY SCIENCE James Yeates
THE VIKINGS Julian Richards
VIRUSES Dorothy H. Crawford
VOLTAIRE Nicholas Cronk
WAR AND TECHNOLOGY
Alex Roland
WATER John Finney
WEATHER Storm Dunlop
THE WELFARE STATE David Garland
WILLIAM SHAKESPEARE Stanley Wells
WITCHCRAFT Malcolm Gaskill
WITTGENSTEIN A. C. Grayling
WORK Stephen Fineman
WORLD MUSIC Philip Bohlman
THE WORLD TRADE
ORGANIZATION Amrita Narlikar
WORLD WAR II Gerhard L. Weinberg
WRITING AND SCRIPT
Andrew Robinson
ZIONISM Michael Stanislawski

Available soon:

AMERICAN CULTURAL HISTORY
Eric Avila
THE BOOK OF COMMON PRAYER
Brian Cummings
MODERN ARCHITECTURE
Adam Sharr
SEXUAL SELECTION Marlene Zuk and
Leigh W. Simmons
THE SAINTS Simon Yarrow

For more information visit our website

www.oup.com/vsi/

Margaret A. Boden

ARTIFICIAL INTELLIGENCE

A Very Short Introduction

OXFORD
UNIVERSITY PRESS

Great Clarendon Street, Oxford, OX2 6DP,
United Kingdom

Oxford University Press is a department of the University of Oxford.
It furthers the University's objective of excellence in research, scholarship,
and education by publishing worldwide. Oxford is a registered trade mark of
Oxford University Press in the UK and in certain other countries

First published in hardback as *AI: Its Nature and Future* 2016
First published as a *Very Short Introduction* 2018

Published in the United States of America by Oxford University Press
198 Madison Avenue, New York, NY 10016, United States of America

British Library Cataloguing in Publication Data
Data available

Library of Congress Control Number: 2018940942

ISBN 978–0–19–960291–9

Printed and bound by
CPI Group (UK) Ltd, Croydon, CR0 4YY

For Byron, Oscar, Lukas, and Alina

Contents

Acknowledgements

I thank the following friends for their very helpful advice (any mistakes, of course, are mine): Phil Husbands, Jeremy Reffin, Anil Seth, Aaron Sloman, and Blay Whitby. And I thank Latha Menon for her understanding and patience.

List of illustrations

Chapter 1
What is artificial intelligence?

Artificial intelligence (AI) seeks to make computers do the sorts of things that minds can do.

Some of these (e.g. reasoning) are normally described as 'intelligent'. Others (e.g. vision) aren't. But all involve psychological skills—such as perception, association, prediction, planning, motor control—that enable humans and animals to attain their goals.

Intelligence isn't a single dimension, but a richly structured space of diverse information-processing capacities. Accordingly, AI uses many different techniques, addressing many different tasks.

And it's everywhere.

AI's practical applications are found in the home, the car (and the driverless car), the office, the bank, the hospital, the sky... and the Internet, including the Internet of Things (which connects the ever-multiplying physical sensors in our gadgets, clothes, and environments). Some lie outside our planet: robots sent to the Moon and Mars, or satellites orbiting in space. Hollywood animations, video and computer games, sat-nav systems, and Google's search engine are all based on AI techniques. So are the systems used by financiers to predict movements on the stock market, and by national governments to help guide policy decisions in health and transport. So are the apps on mobile phones. Add avatars in virtual reality, and the toe-in-the-water models of emotion developed for 'companion' robots. Even art galleries use AI—on their websites, and also in exhibitions of computer art. Less happily, military drones roam today's battlefields—but, thankfully, robot minesweepers do so too.

AI has two main aims. One is *technological*: using computers to get useful things done (sometimes by employing methods very *unlike* those used by minds). The other is *scientific*: using AI concepts and models to help answer questions about human beings and other living things. Most AI workers focus on only one of these, but some consider both.

Besides providing countless technological gizmos, AI has deeply influenced the life sciences.

In particular, AI has enabled psychologists and neuroscientists to develop powerful theories of the mind-brain. These include models of *how the physical brain works*, and—a different, but equally important, question—*just what it is that the brain is doing*: what computational (psychological) questions it is answering, and what sorts of information processing enable it to do that. Many unanswered questions remain, for AI itself has taught us that our minds are very much richer than psychologists had previously imagined.

Biologists, too, have used AI—in the form of 'artificial life' (A-Life), which develops computer models of differing aspects of living organisms. This helps them to explain various types of animal behaviour, the development of bodily form, biological evolution, and the nature of life itself.

Besides affecting the life sciences, AI has influenced philosophy. Many philosophers today base their accounts of mind on AI concepts. They use these to address, for instance, the notorious mind–body problem, the conundrum of free will, and the many puzzles regarding consciousness. However, these philosophical ideas are hugely controversial. And there are deep disagreements about whether any AI system could possess *real* intelligence, creativity, or life.

Last, but not least, AI has challenged the ways in which we think about humanity—and its future. Indeed, some people worry about whether we actually have a future, because they foresee AI surpassing human intelligence across the board. Although a few thinkers welcome this prospect, most dread it: what place will remain, they ask, for human dignity and responsibility?

All these issues are explored in the following chapters.

Virtual machines

'To think about AI', someone might say, 'is to think about computers'. Well, yes and no. The computers, as such, aren't the point. It's what they *do* that matters. In other words, although AI needs *physical* machines (i.e. computers), it's best thought of as using what computer scientists call *virtual* machines.

A virtual machine isn't a machine depicted in virtual reality, nor something like a simulated car engine used to train mechanics. Rather, it's the *information-processing system* that the programmer has in mind when writing a program, and that people have in mind when using it.

A word processor, for example, is thought of by its designer, and experienced by its users, as dealing directly with words and paragraphs. But the program itself usually contains neither. And a neural network (see Chapter 4) is thought of as doing information processing *in parallel*, even though it's usually implemented in a (sequential) von Neumann computer.

That's not to say that a virtual machine is just a convenient fiction, a thing merely of our imagination. Virtual machines are actual realities. They can make things happen, both inside the system and (if linked to physical devices such as cameras or robot hands) in the outside world. AI workers trying to discover what's going wrong when a program does something unexpected only rarely consider hardware faults. Usually, they're interested in the events and causal interactions in the *virtual* machinery, or software.

Programming languages, too, are virtual machines (whose instructions have to be translated into machine code before they can be run). Some are defined in terms of lower-level programming languages, so translation is required at several levels.

That's not true only of programming languages. Virtual machines in general are comprised of patterns of activity (information processing) that exist at various levels. Moreover, it's not true only of virtual machines running on computers. We'll see in Chapter 6 that *the human mind* can be understood as the virtual machine—or rather, the set of mutually interacting virtual machines, running in parallel (and developed or learned at different times)—that is implemented in the brain.

Progress in AI requires progress in defining interesting/useful virtual machines. More *physically* powerful computers (larger, faster) are all very well. They may even be necessary for certain kinds of virtual machines to be implemented. But they can't be exploited unless *informationally* powerful virtual machines can be run on them. (Similarly, progress in neuroscience requires better understanding of what *psychological* virtual machines are being implemented by the physical neurons: see Chapter 7.)

Different sorts of external-world information are used. Every AI system needs input and output devices, if only a keyboard and a screen. Often, there are also special-purpose sensors (perhaps cameras, or pressure-sensitive whiskers) and/or effectors (perhaps sound synthesizers for music or speech, or robot hands). The AI program connects with—causes changes in—these computer-world interfaces as well as processing information internally.

AI processing usually also involves *internal* input and output devices, enabling the various virtual machines within the whole system to interact with each other. For example, one part of a chess program may detect a possible threat by noticing something happening in another, and may then interface with yet another in searching for a blocking move.

The major types of AI

How the information is processed depends on the virtual machine involved. As we'll see in later chapters, there are five major types, each including many variations. One is classical, or symbolic, AI—sometimes called GOFAI (Good Old-Fashioned AI). Another is artificial neural networks, or connectionism. In addition, there are evolutionary programming; cellular automata; and dynamical systems.

Individual researchers often use only one method, but *hybrid* virtual machines also occur. For instance, a theory of human action that switches continually between symbolic and connectionist processing is mentioned in Chapter 4. (This explains why, and how, it is that someone may be distracted from following through on a planned task by noticing something unrelated to it in the environment.) And a sensorimotor device that combines 'situated' robotics, neural networks, and evolutionary programming is described in Chapter 5. (This device helps a robot to find its way 'home' by using a cardboard triangle as a landmark.)

Besides their practical applications, these approaches can illuminate mind, behaviour, and life. Neural networks are helpful for modelling aspects of the brain, and for doing pattern recognition and learning. Classical AI (especially when combined with statistics) can model learning too, and also planning and reasoning. Evolutionary programming throws light on biological evolution and brain development. Cellular automata and dynamical systems can be used to model development in living organisms. Some methodologies are closer to biology than to psychology, and some are closer to non-reflective behaviour than to deliberative thought. To understand the full range of mentality, all of them will be needed—and probably more.

Many AI researchers don't care about how minds work: they seek technological efficiency, not scientific understanding. Even if their techniques originated in psychology, they now bear scant relation to it. We'll see, however, that progress in general-purpose AI (artificial general intelligence, or AGI) will require deep understanding of the computational architecture of minds.

AI foreseen

AI was foreseen in the 1840s by Lady Ada Lovelace. More accurately, she foresaw *part* of it. She focused on symbols and logic, having no glimmering of neural networks, or of evolutionary and dynamical AI. Nor did she have any leanings towards AI's psychological aim, her interest being purely technological.

She said, for instance, that a machine 'might compose elaborate and scientific pieces of music of any degree of complexity or extent', and might also express 'the great facts of the natural world' in enabling 'a glorious epoch in the history of the sciences'. (So she wouldn't have been surprised to see that, two centuries later, scientists are using 'Big Data' and specially crafted programming tricks to advance knowledge in genetics, pharmacology, epidemiology . . . the list is endless.)

The machine she had in mind was the Analytical Engine. This gears-and-cogwheels device (never fully built) had been designed by her close friend Charles Babbage in 1834. Despite being dedicated to algebra and numbers, it was essentially equivalent to a general-purpose digital computer.

Ada Lovelace recognized the potential generality of the Engine, its ability to process symbols representing 'all subjects in the universe'. She also described various basics of modern programming: stored programs, hierarchically nested subroutines, addressing, microprogramming, looping, conditionals, comments, and even bugs. But she said nothing about *just how* musical composition, or scientific reasoning, could be implemented on Babbage's machine. AI was possible, yes—but how to achieve it was still a mystery.

How AI began

That mystery was clarified a century later by Alan Turing. In 1936, Turing showed that every possible computation can in principle be performed by a mathematical system now called a universal Turing machine. This imaginary system builds, and modifies, combinations of binary symbols—represented as '0' and '1'. After codebreaking at Bletchley Park during World War II, he spent the rest of the 1940s thinking about how the abstractly defined Turing machine could be approximated by a physical machine (he helped design the first modern computer, completed in Manchester in 1948), and how such a contraption could be induced to perform intelligently.

Unlike Ada Lovelace, Turing accepted both goals of AI. He wanted the new machines to do useful things normally said to require intelligence (perhaps by using highly unnatural techniques), and also to model the processes occurring in biologically based minds.

The 1950 paper in which he jokily proposed the Turing Test (see Chapter 6) was primarily intended as a manifesto for AI. (A fuller version had been written soon after the war, but the Official Secrets Act prevented publication.) It identified key questions about the information processing involved in intelligence (game playing, perception, language, and learning), giving tantalizing hints about what had already been achieved. (Only 'hints', because the work at Bletchley Park was still top-secret.) It even suggested computational approaches—such as neural networks and evolutionary computing—that became prominent only much later. But the mystery was still far from dispelled. These were highly general remarks: programmatic, not programs.

Turing's conviction that AI must be somehow possible was bolstered in the early 1940s by the neurologist/psychiatrist Warren McCulloch and the mathematician Walter Pitts. In their paper 'A Logical Calculus of the Ideas Immanent in Nervous Activity', they united Turing's work with two other exciting items (both dating from the early 20th century): Bertrand Russell's propositional logic and Charles Sherrington's theory of neural synapses.

The key point about propositional logic is that it's binary. Every sentence (also called a *proposition*) is assumed to be either *true* or *false*. There's no middle way, no recognition of uncertainty or probability. Only two 'truth-values' are allowed, namely *true* and *false*.

Moreover, complex propositions are built, and deductive arguments are carried out, by using logical operators (such as *and*, *or*, and *if–then*) whose meanings are defined in terms of the truth/falsity of the component propositions. For instance, if two (or more) propositions are linked by *and*, it's assumed that both/all of them are true. So 'Mary married Tom and Flossie married Peter' is true if, and only if, *both* 'Mary married Tom' and 'Flossie married Peter' are true.

Russell and Sherrington could be brought together by McCulloch and Pitts because they had both described binary systems. The *true/false* values of logic were mapped onto the *on/off* activity of brain cells and the *0/1* of individual states in Turing machines. Neurons were believed by Sherrington to be not only strictly on/off, but also to have fixed thresholds. So logic gates (computing *and*, *or*, and *not*) were defined as tiny neural nets, which could be interconnected to represent highly complex propositions. Anything that could be stated in propositional logic could be computed by some neural network, and by some Turing machine.

In brief, neurophysiology, logic, and computation were bundled together—and psychology came along too. McCulloch and Pitts believed (as many philosophers then did) that natural language boils down, in essence, to logic. So all reasoning and opinion, from scientific argument to schizophrenic delusions, was grist for their theoretical mill. They foresaw a time when, for the whole of psychology, 'specification of the [neural] net would contribute all that could be achieved in that field'.

The core implication was clear: *one and the same theoretical approach*—namely, Turing computation—could be applied to human and machine intelligence.

Turing, of course, agreed. But he couldn't take AI much further: the technology available was too primitive. In the mid-1950s, however, more powerful and/or easily usable machines were developed. 'Easily usable', here, means that it was easier to define new *virtual* machines (e.g. programming languages), which could be more easily used to define higher-level virtual machines (e.g. programs to do mathematics, or planning).

Symbolic AI research, broadly in the spirit of Turing's manifesto, commenced on both sides of the Atlantic. One late-1950s landmark was Arthur Samuel's checkers (draughts) player, which made newspaper headlines because it learned to beat Samuel himself. That was an intimation that computers might one day develop *super*human intelligence, outstripping the capacities of their programmers.

The second such intimation also occurred in the late 1950s, when the Logic Theory Machine not only proved eighteen of Russell's key logical theorems, but found a more elegant proof of one of them. This was truly impressive. Whereas Samuel was only a mediocre checkers player, Russell was a world-leading logician. (Russell himself was delighted by this achievement, but the *Journal of Symbolic Logic* refused to publish a paper with a computer program named as an author, especially as it hadn't proved a *new* theorem.)

The Logic Theory Machine was soon outdone by the General Problem Solver (GPS)—'outdone' not in the sense that GPS could surpass yet more towering geniuses, but in the sense that it wasn't limited to only one field. As the name suggests, GPS could be applied to any problem that could be represented (as explained in Chapter 2) in terms of goals, sub-goals, actions, and operators. It was up to the programmers to identify the goals, actions, and operators relevant for any specific field. But once that had been done, the *reasoning* could be left to the program.

GPS managed to solve the 'missionaries-and-cannibals' problem, for example. (*Three missionaries and three cannibals on one side of a river; a boat big enough for two people; how can everyone cross the river, without cannibals ever outnumbering missionaries?*) That's difficult even for humans, because it requires one to go backwards in order to go forwards. (Try it, using pennies!)

The Logic Theory Machine and GPS were early examples of GOFAI. They are now 'old-fashioned', to be sure. But they were also 'good', for they pioneered the use of *heuristics* and *planning*—both of which are hugely important in AI today (see Chapter 2).

GOFAI wasn't the only type of AI to be inspired by the 'Logical Calculus' paper. Connectionism, too, was encouraged by it. In the 1950s, networks of McCulloch–Pitts logical neurons, either purpose-built or simulated on digital computers, were used (by Albert Uttley, for instance) to model associative learning and conditioned reflexes. (Unlike today's neural networks, these did *local*, not *distributed*, processing: see Chapter 4.)

But early network modelling wasn't wholly dominated by neuro-logic. The systems implemented (in analogue computers) by Raymond Beurle in the mid-1950s were very different. Instead of carefully designed networks of logic gates, he started from two-dimensional (2D) arrays of randomly connected, and varying-threshold, units. He saw neural self-organization as due to dynamical waves of activation—building, spreading, persisting, dying, and sometimes interacting.

As Beurle realized, to say that psychological processes could be *modelled by* a logic-chopping machine wasn't to say that the brain *actually is* such a machine. McCulloch and Pitts had already pointed this out. Only four years after their first groundbreaking paper, they had published another one arguing that thermodynamics is closer than logic to the functioning of the brain. Logic gave way to statistics, single units to collectivities, and deterministic purity to probabilistic noise.

In other words, they had described what's now called distributed, error-tolerant computing (see Chapter 4). They saw this new approach as an 'extension' of their previous one, not a contradiction of it. But it was more biologically realistic.

Cybernetics

McCulloch's influence on early AI went even further than GOFAI and connectionism. His knowledge of neurology as well as logic made him an inspiring leader in the budding cybernetics movement of the 1940s.

The cyberneticians focused on biological self-organization. This covered various kinds of adaptation and metabolism, including autonomous thought and motor behaviour as well as (neuro) physiological regulation. Their central concept was 'circular causation', or feedback. And a key concern was teleology, or purposiveness. These ideas were closely related, for feedback depended on goal differences: the current distance from the goal was used to guide the next step.

Norbert Wiener (who designed anti-ballistic missiles during the war) named the movement in 1948, defining it as 'the study of control and communication in the animal and the machine'. Those cyberneticians who did computer modelling often drew inspiration from control engineering and analogue computers rather than logic and digital computing. However, the distinction wasn't clear-cut. For instance, goal differences were used both to control guided missiles and to direct symbolic problem solving. Moreover, Turing—the champion of classical AI—used dynamical equations (describing chemical diffusion) to define self-organizing systems in which novel structure, such as spots or segmentation, could emerge from a homogeneous origin (see Chapter 5).

Other early members of the movement included the experimental psychologist Kenneth Craik; the mathematician John von Neumann; the neurologists William Grey Walter and William Ross Ashby; the engineer Oliver Selfridge; the psychiatrist and anthropologist Gregory Bateson; and the chemist and psychologist Gordon Pask.

Craik, who died (aged 31) in a cycling accident in 1943, before the advent of digital computers, referred to analogue computing in thinking about the nervous system. He described perception and motor action, and intelligence in general, as guided by feedback from 'models' in the brain. His concept of cerebral models, or representations, would later be hugely influential in AI.

Von Neumann had puzzled about self-organization throughout the 1930s, and was hugely excited by McCulloch and Pitts' first paper. Besides changing his basic computer design from decimal to binary, he adapted their ideas to explain biological evolution and reproduction. He defined various cellular automata: systems made of many computational units, whose changes follow simple rules depending on the current state of neighbouring units. Some of these could replicate others. He even defined a universal replicator, capable of copying anything—itself included. Replication errors, he said, could lead to evolution.

Cellular automata were specified by von Neumann in abstract informational terms. But they could be embodied in many ways, for example, as self-assembling robots, Turing's chemical diffusion, Beurle's physical waves, or—as soon became clear—DNA.

From the late 1940s on, Ashby developed the Homeostat, an electrochemical model of physiological homeostasis. This intriguing machine could settle into an overall equilibrium state no matter what values were initially assigned to its hundred parameters (allowing almost 400,000 different starting conditions). It illustrated Ashby's theory of dynamical adaptation—both inside the body (not least, the brain) and between the body and its external environment, in trial-and-error learning and adaptive behaviour.

Grey Walter, too, was studying adaptive behaviour—but in a very different way. He built mini-robots resembling tortoises, whose sensorimotor circuitry modelled Sherrington's theory of neural reflexes. These pioneering situated robots displayed lifelike behaviours such as light-seeking, obstacle-avoidance, and associative learning via conditioned reflexes. They were exhibited to the general public at the Festival of Britain in 1951.

Ten years later, Selfridge (grandson of the founder of the London department store) used symbolic methods to implement an essentially parallel-processing system called Pandemonium.

This GOFAI program learned to recognize patterns by having many bottom-level 'demons', each always looking out for one simple perceptual input, which passed their results on to higher-level demons. These weighed the features recognized so far for consistency (e.g. only *two* horizontal bars in an **F**), downplaying any features that didn't fit. Confidence levels could vary, and they mattered: the demons that shouted loudest had the greatest effect. Finally, a master-demon chose the most plausible pattern, given the (often conflicting) evidence available. This research soon influenced both connectionism and symbolic AI. (One very recent offshoot is the LIDA model of consciousness: see Chapter 6.)

Bateson had little interest in machines, but he based his 1960s theories of culture, alcoholism, and 'double-bind' schizophrenia on ideas about communication (i.e. feedback) picked up earlier at cybernetic meetings. And from the mid-1950s on, Pask—described as 'the genius of self-organizing systems' by McCulloch—used cybernetic and symbolic ideas in many different projects. These included interactive theatre; intercommunicating musical robots; architecture that learned and adapted to its users' goals; chemically self-organizing concepts; and teaching machines. The latter enabled people to take different routes through a complex knowledge representation, so were suitable for both step-by-step and holistic cognitive styles (and varying tolerance of irrelevance) on the learner's part.

In brief, all the main types of AI were being thought about, and even implemented, by the late 1960s—and in some cases, much earlier than that.

Most of the researchers concerned are widely revered today. But only Turing was a constant spectre at the AI feast. For many years, the others were remembered only by some subset of the research community. Grey Walter and Ashby, in particular, were nearly forgotten until the late 1980s, when they were lauded (alongside Turing) as grandfathers of A-Life. To understand why, one must know how the computer modellers became disunited.

How AI divided

Before the 1960s, there was no clear distinction between people modelling language or logical thinking and people modelling purposive/adaptive motor behaviour. Some individuals worked on both. (Donald Mackay even suggested building hybrid computers, combining neural networks with symbolic processing.) And all were mutually sympathetic. Researchers studying physiological self-regulation saw themselves as engaged in the same overall enterprise as their psychologically oriented colleagues. They all attended the same meetings: the interdisciplinary Macy seminars in the USA (chaired by McCulloch from 1946 to 1951), and London's seminal conference on 'The Mechanization of Thought Processes' (organized by Uttley in 1958).

From about 1960, however, an intellectual schism developed. Broadly speaking, those interested in *life* stayed in cybernetics, and those interested in *mind* turned to symbolic computing. The network enthusiasts were interested in both brain and mind, of course. But they studied associative learning in general, not specific semantic content or reasoning, so fell within cybernetics rather than symbolic AI. Unfortunately, there was scant mutual respect between these increasingly separate sub-groups.

The emergence of distinct sociological coteries was inevitable. For the theoretical questions being asked—biological (of varying kinds) and psychological (also of varying kinds)—were different. So too were the technical skills involved: broadly defined, logic versus differential equations. Growing specialization made communication increasingly difficult, and largely unprofitable. Highly eclectic conferences became a thing of the past.

Even so, the division needn't have been so ill-tempered. The bad feeling on the cybernetic/connectionist side began as a mixture of professional jealousy and righteous indignation. These were prompted by the huge initial success of symbolic computing, by the journalistic interest attending the provocative term 'artificial intelligence' (coined by John McCarthy in 1956 to name what had previously been called 'computer simulation'), and by the arrogance—and unrealistic hype—expressed by some of the symbolists.

Members of the symbolist camp were initially less hostile, because they saw themselves as winning the AI competition. Indeed, they largely ignored the early network research, even though some of their leaders (Marvin Minsky, for instance) had started out in that area.

In 1958, however, an ambitious theory of neurodynamics—defining parallel-processing systems capable of self-organized learning from a random base (and error-tolerant to boot)—was presented by Frank Rosenblatt and partially implemented in his photoelectric Perceptron machine. Unlike Pandemonium, this didn't need the input patterns to be pre-analysed by the programmer. This novel form of connectionism couldn't be ignored by the symbolists. But it was soon contemptuously dismissed. As explained in Chapter 4, Minsky (with Seymour Papert) launched a stinging critique in the 1960s claiming that perceptrons are incapable of computing some basic things.

Funding for neural-network research dried up accordingly. This outcome, deliberately intended by the two attackers, deepened the antagonisms within AI.

To the general public, it now seemed that classical AI was the only game in town. Admittedly, Grey Walter's tortoises had received great acclaim in the Festival of Britain. Rosenblatt's Perceptron was hyped by the press in the late 1950s, as was Bernard Widrow's pattern-learning Adaline (based on signal-processing). But the symbolists' critique killed that interest stone dead. It was symbolic

AI which dominated the media in the 1960s and 1970s (and which influenced the philosophy of mind as well).

That situation didn't last. Neural networks—as 'PDP systems' (doing parallel distributed processing)—burst onto the public stage again in 1986 (see Chapter 4). Most outsiders—and some insiders, who should have known better—thought of this approach as utterly *new*. It seduced the graduate students, and attracted enormous journalistic (and philosophical) attention. Now, it was the symbolic AI people whose noses were put out of joint. PDP was in fashion, and classical AI was widely said to have failed.

As for the other cyberneticians, they finally came in from the cold with the naming of A-Life in 1987. The journalists, and the graduate students, followed. Symbolic AI was challenged yet again.

In the 21st century, however, it has become clear that different questions require different types of answers—horses for courses. Although traces of the old animosities remain, there's now room for respect, and even cooperation, between different approaches. For instance, 'deep learning' is sometimes used in powerful systems combining symbolic logic with multilayer probabilistic networks; and other hybrid approaches include ambitious models of consciousness (see Chapter 6).

Given the rich variety of virtual machines that constitute the human mind, one shouldn't be too surprised.

Chapter 2
General intelligence as the Holy Grail

State-of-the-art AI is a many-splendoured thing. It offers a profusion of virtual machines, doing many different kinds of information processing. There's no key secret here, no core technique unifying the field: AI practitioners work in highly diverse areas, sharing little in terms of goals and methods. This book can mention only very few of the recent advances. In short, AI's methodological range is extraordinarily wide.

One could say that it's been astonishingly successful. For its practical range, too, is extraordinarily wide. A host of AI applications exist, designed for countless specific tasks and used in almost every area of life, by laymen and professionals alike. Many outperform even the most expert humans. In that sense, progress has been spectacular.

But the AI pioneers weren't aiming only for specialist systems. They were also hoping for systems with *general* intelligence. Each human-like capacity they modelled—vision, reasoning, language, learning, and so on—would cover its entire range of challenges. Moreover, these capacities would be integrated when appropriate.

Judged by those criteria, progress has been far less impressive. John McCarthy recognized AI's need for 'common sense' very early on. And he spoke on 'Generality in Artificial Intelligence' in both of his high-visibility Turing Award addresses, in 1971 and 1987—but he was complaining, not celebrating. In 2018, his complaints aren't yet answered.

The 21st century is seeing a revival of interest in AGI, driven by recent increases in computer power. If that were achieved, AI systems could rely less on special-purpose programming tricks, benefitting instead from general powers of reasoning and perception—plus language, creativity, and emotion (all of which are discussed in Chapter 3).

However, that's easier said than done. General intelligence is still a major challenge, still highly elusive. AGI is the field's Holy Grail.

Supercomputers aren't enough

Today's supercomputers are certainly a help to anyone seeking to realize this dream. The combinatorial explosion—wherein more computations are required than can actually be executed—is no longer the constant threat that it used to be. Nevertheless, problems can't always be solved merely by increasing computer power.

New problem-solving *methods* are often needed. Moreover, even if a particular method *must* succeed in principle, it may need too much time and/or memory to succeed in practice. Three such examples (concerning neural networks) are given in Chapter 4.

Efficiency is important, too: the fewer the number of computations, the better. In short, problems must be made tractable.

There are several basic strategies for doing that. All were pioneered by classical symbolic AI, or GOFAI, and all are still essential today.

One is to direct attention to only a part of the *search space* (the computer's representation of the problem, within which the solution is assumed to be located). Another is to construct a smaller search space by making simplifying assumptions. A third is to order the search efficiently. Yet another is to construct a different search space, by representing the problem in a new way.

These approaches involve *heuristics*, *planning*, *mathematical simplification*, and *knowledge representation*, respectively. The next five sections consider those general AI strategies.

Heuristic search

The word 'heuristic' has the same root as '*Eureka!*': it comes from the Greek for *find*, or *discover*. Heuristics were highlighted by early GOFAI, and are often thought of as 'programming tricks'. But the term didn't originate with programming: it has long been familiar to logicians and mathematicians.

Whether in humans or machines, heuristics make it easier to solve the problem. In AI, they do this by directing the program towards certain parts of the search space and away from others.

Many heuristics, including most of those used in the very early days of AI, are rules of thumb that aren't guaranteed to succeed. The solution may lie in some part of the search space that the heuristic has led the system to ignore. For example, 'Protect your queen' is a very helpful rule in chess, but it should occasionally be disobeyed.

Others can be logically or mathematically proved to be adequate. Much work in AI and computer science today aims to identify provable properties of programs. That's one aspect of 'Friendly AI', because human safety may be jeopardized by the use of logically unreliable systems (see Chapter 7).

Whether reliable or not, heuristics are an essential aspect of AI research. The increasing AI specialism mentioned earlier depends partly on the definition of new heuristics that can improve efficiency spectacularly, but only in one highly restricted sort of problem, or search space. A hugely successful heuristic may not be suitable for 'borrowing' by other AI programs.

Given several heuristics, their order of application may matter. For instance, 'Protect your queen' should be taken into account before 'Protect your bishop'—even though this ordering will occasionally lead to disaster. Different orderings will define different search trees through the search space. Defining and ordering heuristics are crucial tasks for modern AI. (Heuristics are prominent in cognitive psychology, too. Intriguing work on 'fast and frugal heuristics', for example, indicates how evolution has equipped us with efficient ways of responding to the environment.)

Heuristics make brute-force search through the entire search space unnecessary. But they are sometimes combined with (limited) brute-force search. IBM's chess program *Deep Blue*, which caused worldwide excitement by beating world champion Gary Kasparov in 1997, used dedicated hardware chips, processing 200 million positions per second, to generate every possible move for the next eight.

However, it had to use heuristics to select the 'best' move within them. And since its heuristics weren't reliable, even *Deep Blue* didn't beat Kasparov *every* time.

Planning

Planning, too, is prominent in today's AI—not least in a wide range of military activities. Indeed, the USA's Department of Defense, which paid for the majority of AI research until very recently, has said that the money saved (by AI planning) on battlefield logistics in the first Iraq war outweighed all their previous investment.

Planning isn't restricted to AI: we all do it. Think of packing for your holiday, for instance. You have to find all the things you want to take, which probably won't all be found in the same place. You may have to buy some new items (sun cream, perhaps). You must decide whether to collect all the things together (perhaps on your bed, or on a table), or whether to put each one in your luggage when you find it. That decision will depend in part on whether you want to put the clothes in last of all, to prevent creasing. You'll need a rucksack, or a suitcase, or maybe two: how do you decide?

The GOFAI programmers who used planning as an AI technique had such consciously thought-out examples in mind. That's because the pioneers responsible for the Logic Theory Machine (see Chapter 1) and GPS were primarily interested in the psychology of human reasoning.

Modern AI planners don't rely so heavily on ideas garnered from conscious introspection or experimental observation. And their plans are very much more complex than was possible in the early days. But the basic idea is the same.

A plan specifies a sequence of actions, represented at a general level—a final goal, plus sub-goals and sub-sub-goals...—so that all details aren't considered at once. Planning at a suitable level of abstraction can lead to tree pruning within the search space, so some details never need to be considered at all. Sometimes, the final goal is *itself* a plan of action—perhaps scheduling the deliveries to and from a factory or battlefield. At other times, it's the answer to a question—for example, a medical diagnosis.

For any given goal, and expected situations, the planning program needs: a list of actions—that is, symbolic operators—or action types (instantiated by filling in parameters derived from the problem), each of which can make some relevant change; for every action, a set of necessary prerequisites (cf. to grasp something, it must be within reach); and heuristics for prioritizing the required changes and ordering the actions. If the program decides on a particular action, it may have to set up a new sub-goal to satisfy the prerequisites. This goal-formulating process can be repeated again and again.

Planning enables the program—and/or the human user—to discover what actions have already been taken, and why. The 'why' refers to the goal hierarchy: *this* action was taken to satisfy *that* prerequisite, to achieve *such-and-such* a sub-goal. AI systems commonly employ techniques of 'forward-chaining' and 'backward-chaining', which explain how the program found its solution. This helps the user to judge whether the action/advice of the program is appropriate.

Some current planners have tens of thousands of lines of code, defining hierarchical search spaces on numerous levels. These systems are often significantly different from the early planners.

For example, most don't assume that all the sub-goals can be worked on independently (i.e. that problems are *perfectly decomposable*). In real life, after all, the result of one goal-directed activity may be undone by another. Today's planners can handle *partially decomposable* problems: they work on sub-goals independently, but can do extra processing to combine the resulting sub-plans if necessary.

The classical planners could tackle only problems in which the environment was fully observable, deterministic, finite, and static. But some modern planners can cope with environments that are partially observable (i.e. the system's model of the world may be incomplete and/or incorrect) and probabilistic. In those cases, the system must monitor the changing situation during execution, so as to make changes in the plan—and/or in its own 'beliefs' about the world—as appropriate. Some modern planners can do this over very long periods: they engage in continuous goal formulation, execution, adjustment, and abandonment, according to the changing environment.

Many other developments have been added, and are still being added, to classical planning. It may seem surprising, then, that planning was roundly rejected by some roboticists in the 1980s, 'situated' robotics being recommended instead (see Chapter 5). The notion of internal representation—of goals and possible actions, for example—was rejected as well. However, that criticism was largely mistaken. Robotics often needs planning as well as purely reactive responses—to build soccer-playing robots, for instance.

Mathematical simplification

Whereas heuristics leave the search space as it is (making the program focus on only part of it), simplifying assumptions construct an unrealistic—but computationally tractable—search space.

Some such assumptions are mathematical. One example is the 'i.i.d.' assumption, commonly used in machine learning. This represents the probabilities in the data as being much simpler than they actually are.

The advantage of mathematical simplification when defining the search space is that mathematical—that is, clearly definable and, to mathematicians at least, readily intelligible—methods of search can be used. But that's not to say that *any* mathematically defined search will be useful. As noted earlier, a method that's mathematically *guaranteed* to solve every problem within a certain class may be unusable in real life, because it would need infinite time to do so. It may, however, suggest approximations that are more practicable: see the discussion of 'backprop' in Chapter 4.

Non-mathematical simplifying assumptions in AI are legion—and often unspoken. One is the (tacit) assumption that problems can be defined and solved without taking emotions into account (see Chapter 3). Many others are built into the general knowledge representation that's used in specifying the task.

Knowledge representation

Often, the hardest part of AI problem solving is presenting the problem to the system in the first place. Even if it *seems* that someone can communicate directly with a program—by speaking in English to *Siri*, perhaps, or by typing French words into Google's search engine—they can't. Whether one's dealing with texts or with images, the information ('knowledge') concerned must be presented to the system in a fashion that the machine can understand—in other words, that it can deal with. (Whether that is *real* understanding is discussed in Chapter 6.)

AI's ways of doing this are highly diverse. Some are developments/variations of general methods of knowledge representation introduced in GOFAI. Others, increasingly, are highly specialized methods, tailor-made for a narrow class of problems. There may be, for instance, a new way of representing X-ray images, or photographs of a certain class of cancerous cells, carefully tailored to enable some highly specific method of medical interpretation (so, no good for recognizing cats, or even CAT scans).

In the quest for AGI, the *general* methods are paramount. Initially inspired by psychological research on human cognition, these include: sets of IF–THEN rules; representations of individual concepts; stereotyped action sequences; semantic networks; and inference by logic or probability.

Let's consider each of these in turn. (Another form of knowledge representation, namely neural networks, is described in Chapter 4.)

Rule-based programs

In rule-based programming, a body of knowledge/belief is represented as a set of IF–THEN rules linking Conditions to Actions: IF this Condition is satisfied, THEN take that Action. This form of knowledge representation draws on formal logic (Emil Post's 'production' systems). But the AI pioneers Allen Newell and Herbert Simon believed it to underlie human psychology in general.

Both Condition and Action may be complex, specifying a conjunction (or disjunction) of several—perhaps many—items. If several Conditions are satisfied simultaneously, the most inclusive conjunction is given priority. So 'IF the goal is to cook roast beef and Yorkshire pudding' will take precedence over 'IF the goal is to cook roast beef'—and adding 'and three veg' to the Condition will trump that.

Rule-based programs don't specify the order of steps in advance. Rather, each rule lies in wait to be triggered by its Condition. Nevertheless, such systems can be used to do planning. If they couldn't, they would be of limited use for AI. But they do it differently from how it's done in the oldest, most familiar, form of programming (sometimes called 'executive control').

In programs with executive control planning is represented explicitly. The programmer specifies a sequence of goal-seeking instructions to be followed step by step, in strict temporal order: '*Do this*, then *do that*; then *look to see* whether *X* is true; if it is, *do such-and-such*; if not, *do so-and-so*'.

Sometimes, the '*this*' or the '*so-and-so*' is an explicit instruction to set up a goal or sub-goal. For instance, a robot with the goal of leaving the room may be instructed to set up the sub-goal of opening the door; next, if examining the current state of the door shows it to be closed, set up the sub-sub-goal of grasping the door handle. (A human toddler may need a sub-sub-sub-goal—namely, getting an adult to grasp the unreachable door handle; and the infant may need several goals at even lower levels in order to do that.)

A rule-based program, too, could work out how to escape from the room. However, the plan hierarchy would be represented not as a temporally ordered sequence of explicit steps, but as the logical structure *implicit* in the collection of IF–THEN rules that comprise the system. A Condition may require that such-and-such a goal has already been set up (IF you want to open the door, and you aren't tall enough). Similarly, an Action can include the setting up of a new goal or sub-goal (THEN ask an adult). Lower levels will be activated automatically (IF you want to ask someone to do something, THEN set up the goal of moving near to them).

Of course, the programmer has to have included the relevant IF–THEN rules (in our example, rules dealing with doors and door handles). But he/she doesn't need to have anticipated all the potential logical implications of those rules. (That's a curse, as well as a blessing, because potential inconsistencies may remain undiscovered for quite a while.)

The active goals/sub-goals are posted on a central 'blackboard', which is accessible to the whole system. The information displayed on the blackboard includes not only activated goals but also perceptual input, and other aspects of current processing. (That idea has influenced a leading neuropsychological theory of consciousness, and an AI model of consciousness based on it: see Chapter 6.)

Rule-based programs were widely used for the pioneering 'expert systems' of the early 1970s. These included MYCIN, which offered advice to human physicians on identifying infectious diseases and on prescribing antibiotic drugs, and DENDRAL, which performed spectral analysis of molecules within a particular area of organic chemistry. MYCIN, for instance, did medical diagnosis by matching symptoms and background bodily properties (Conditions) to diagnostic conclusions and/or suggestions for further tests or medication (Actions). Such programs were AI's first move away from the hope of generalism towards the practice of specialism. And they were the first step towards Ada Lovelace's dream of machine-made science (see Chapter 1).

The rule-based form of knowledge representation enables programs to be built gradually, as the programmer—or perhaps an AGI system itself—learns more about the domain. A new rule can be added at any time. There's no need to rewrite the program from scratch. However, there's a catch. If the new rule isn't logically consistent with the existing ones, the system won't always do what it's supposed to do. It may not even *approximate* what it's supposed to do. When dealing with a small set of rules, such logical conflicts are easily avoided, but larger systems are less transparent.

In the 1970s, the new IF–THEN rules were drawn from ongoing conversations with human experts, asked to explain their decisions. Today, many of the rules don't come from conscious introspection. But they are even more efficient. Modern expert systems (a term rarely used today) range from huge programs used in scientific research and commerce to humble apps on phones. Many outperform their predecessors because they benefit from additional forms of knowledge representation, such as statistics and special-purpose visual recognition, and/or the use of Big Data (see Chapter 4).

These programs can assist, or even replace, human experts in narrowly restricted fields. Now, there are countless examples, used to aid human professionals working in science, medicine, law...and even dress design. (Which isn't entirely good news: see Chapter 7.)

Frames, word-vectors, scripts, semantic nets

Other commonly used methods of knowledge representation concern individual concepts, not entire domains (such as medical diagnosis or dress design).

For instance, one can tell a computer what a room is by specifying a hierarchical data structure (sometimes called a 'frame'). This represents a *room* as having *floor*, *ceiling*, *walls*, *doors*, *windows*, and *furniture* (*bed*, *bath*, *dining table*...). Actual rooms have varying numbers of walls, doors, and windows, so 'slots' in the frame allow specific numbers to be filled in—and provide default assignments too (four walls, one door, one window).

Such data structures can be used by the computer to find analogies, answer questions, engage in a conversation, or write or understand a story. And they're the basis of CYC: an ambitious—some would say vastly overambitious—attempt to represent all human knowledge.

Frames can be misleading, however. Default assignments, for instance, are problematic. (Some rooms have no window, and open-plan rooms have no door.) Worse: what of everyday concepts such as *dropping*, or *spilling*? Symbolic AI represents our common-sense knowledge of 'naïve physics' by constructing frames coding such facts as that a physical object will drop if unsupported. But a helium balloon won't. Allowing explicitly for such cases is a never-ending task.

In some applications using recent techniques for dealing with Big Data, a single concept may be represented as a cluster, or 'cloud', made up of hundreds or thousands of sometimes-associated concepts, with the probabilities of the many paired associations being distinguished: see Chapter 3. Similarly, concepts can now be represented by 'word-vectors' rather than words. Here, semantic features that contribute to, and connect, many different concepts are discovered by the (deep-learning) system, and used to predict the following word—in machine translation, for instance. However, these representations aren't yet as amenable for use in reasoning or conversation, as classical frames.

Some data structures (called 'scripts') denote familiar action sequences. For instance, putting a child to bed often involves tucking them up, reading a story, singing a lullaby, and switching on the night light. Such data structures can be used for question-answering, and also for *suggesting* questions. If a mother omits the night light, questions can arise about *Why?* and *What happened next?* In other words, therein lies the seed of a story. Accordingly, this form of knowledge representation is used for automatic story-writing—and would be needed by 'companion' computers capable of engaging in normal human conversation (see Chapter 3).

An alternative form of knowledge representation for concepts is semantic networks (these are *localist* networks: see Chapter 4). Pioneered by Ross Quillian in the 1960s as models of human associative memory, several extensive examples (e.g. *WordNet*) are now available as public data resources. A semantic network links concepts by semantic relations such as *synonymy*, *antonymy*, *subordination*, *super-ordination*, *part-whole*—and often also by associative linkages assimilating *factual* world-knowledge to semantics (see Chapter 3).

The network may represent words as well as concepts, by adding links coding for *syllables*, *initial letters*, *phonetics*, and *homonyms*. Such a network is used by Kim Binsted's JAPE and Graeme Ritchie's STAND UP, which generate jokes (of nine different types) based on puns, alliteration, and syllable-switching. For example: *Q: What do you call a depressed train? A: A low-comotive*; *Q: What do you get if you mix a sheep with a kangaroo? A: A woolly jumper.*

A *caveat:* semantic networks aren't the same thing as neural networks. As we'll see in Chapter 4, *distributed* neural networks represent knowledge in a very different way. There, individual concepts are represented not by a single node in a carefully defined associative net, but by the changing pattern of activity across an entire network. Such systems can tolerate conflicting evidence, so aren't bedevilled by the problems of maintaining logical consistency (to be described in the next section). But they can't do precise inference. Nevertheless, they're a sufficiently important type of knowledge representation (and a sufficiently important basis for practical applications) to merit a separate chapter.

Logic and the semantic web

If one's ultimate aim is AGI, logic seems highly appropriate as a knowledge representation. For logic is *generally* applicable. In principle, the same representation (the same logical symbolism) can be used for vision, learning, language, and so on, and for any integration thereof. Moreover, it provides powerful methods of theorem proving to handle the information.

That's why the preferred mode of knowledge representation in early AI was the predicate calculus. This form of logic has more representational power than propositional logic, because it can 'get inside' sentences to express their meaning. For example, consider the sentence 'This shop has a hat to fit everyone'. Predicate calculus can clearly distinguish these three possible meanings: 'For every human individual, there exists in this shop some hat that will fit them'; 'There exists in this shop a hat whose size can be varied so as to fit any human being'; and 'In this shop there exists a hat [presumably folded up!] large enough to fit all human beings simultaneously'.

For many AI researchers, predicate logic is still the preferred approach. CYC's frames, for example, are based in predicate logic.

So are the natural language processing (NLP) representations in compositional semantics (see Chapter 3). Sometimes, predicate logic is extended so as to represent time, cause, or duty/morality. Of course, that depends on someone's having developed those forms of modal logic—which isn't easy.

However, logic has disadvantages, too. One involves the combinatorial explosion. AI's widely used 'resolution' method for logical theorem proving can get bogged down in drawing conclusions that are true but irrelevant. Heuristics exist for guiding, and restricting, the conclusions—and for deciding when to give up (which the Sorcerer's Apprentice couldn't do). But they aren't foolproof.

Another is that resolution theorem proving assumes that *not-not-X* implies *X*. If the domain being reasoned about is completely understood, that's logically correct. But users of programs (such as many expert systems) with built-in resolution often assume that failure to find a contradiction implies that no contradiction exists—so-called 'negation by failure'. Usually, that's a mistake. In real life, there's a big difference between proving that something is false and failing to prove that it's true (think of wondering whether or not your partner is cheating on you).

A third disadvantage is that in classical ('monotonic') logic, once something is proved to be true, it stays true. In practice, that's not always so. One may accept X for good reason (perhaps it was a default assignment, or even a conclusion from careful argument and/or strong evidence), but it can turn out later that X is no longer true—or wasn't true in the first place. If so, one must revise one's beliefs accordingly. Given a logic-based knowledge representation, that's easier said than done. Many researchers, inspired by McCarthy, have tried to develop 'non-monotonic' logics that can tolerate changing truth-values. Similarly, people have defined various 'fuzzy' logics, where a statement can be labelled as *probable/improbable*, or as *unknown*, rather than *true/false*. Even so, no reliable defence against monotonicity has been found.

AI researchers developing logic-based knowledge representation are increasingly seeking the ultimate atoms of knowledge, or meaning, *in general*. They aren't the first: McCarthy and Hayes did so in 'Some Philosophical Problems from an AI Standpoint'. That early paper addressed many familiar puzzles, from free will to counterfactuals. These included questions about the basic ontology of the universe: states, events, properties, changes, actions... *what?*

Unless one is a metaphysician at heart (a rare human passion), why should one care? And why should these arcane questions be 'increasingly' pursued today? Broadly, the answer is that trying to design AGI raises questions about what ontologies the knowledge representation can use. These questions arise also in designing the semantic web.

The semantic web isn't the same as the World Wide Web—which we've had since the 1990s. For the semantic web isn't even state of the art: it's state of the future. If and when it exists, machine-driven associative search will be improved and supplemented by machine understanding. This will enable apps and browsers to access information from anywhere on the Internet, and to integrate different items sensibly in reasoning about questions. That's a tall order. Besides requiring huge engineering advances in hardware and communications infrastructure, this ambitious project (directed by Sir Tim Berners-Lee) needs to deepen the Web-roaming programs' understanding of what they're doing.

Search engines like Google's, and NLP programs in general, can find associations between words and/or texts—but there's no understanding there. Here, this isn't a philosophical point (for that, see Chapter 6), but an empirical one—and a further obstacle to achieving AGI. Despite some seductively deceptive examples—such as WATSON, *Siri*, and machine translation (all discussed in Chapter 3)—today's computers don't grasp the meaning of what they 'read' or 'say'.

Computer vision

Today's computers don't understand visual images as humans do, either. (Again, this is an *empirical* point: whether AGIs could have conscious visual phenomenology is discussed in Chapter 6.)

Since 1980, the various knowledge representations used for AI vision have drawn heavily on psychology—especially the theories of David Marr and James Gibson. Despite such psychological influences, however, current visual programs are gravely limited.

Admittedly, computer vision has achieved remarkable feats: facial recognition with 98 per cent success, for instance. Or reading cursive handwriting. Or noticing someone behaving suspiciously (continually pausing by car doors) in parking lots. Or identifying certain diseased cells, better than human pathologists can. Faced with such successes, one's mind is strongly tempted to boggle.

But the programs (many are neural networks: see Chapter 4) usually have to know exactly what they're looking for: for example, a face *not* upside down, *not* in profile, *not* partly hidden behind something else, and (for 98 per cent success) lit in a particular way.

That word 'usually' is important. In 2012, Google's Research Laboratory integrated 1,000 large (sixteen-core) computers to form a huge neural network, with over a billion connections. Equipped with deep learning, it was presented with ten million random images from YouTube videos. It wasn't told what to look for, and the images weren't labelled. Nevertheless, after three days one unit (one artificial neuron) had learned to respond to images of a cat's face, and another to human faces.

Impressive? Well, yes. Intriguing, too: the researchers were quick to recall the idea of 'grandmother cells' in the brain. Ever since the 1920s, neuroscientists have differed over whether or not these exist. To say that they do is to say that there are cells in the brain (either single neurons or small groups of neurons) that become active when, and only when, a grandmother, or some other specific feature, is perceived. Apparently, something analogous was going on in Google's cat-recognizing network. And although the cats' faces had to be full on and the right way up, they could vary in size, or appear in different positions within the 200 × 200 array. A further study, which trained the system on carefully pre-selected (but unlabelled) images of human faces, *including some in profile*, resulted in a unit that could sometimes—only *sometimes*—discriminate faces turned away from the viewer.

There are now many more—and even more impressive—such achievements. Multilayer networks have already made huge advances in face recognition, and can sometimes find the most salient part of an image and generate a verbal caption (e.g. 'people shopping in an outdoor market') to describe it. The recently initiated *Large Scale Visual Recognition Challenge* is annually increasing the number of visual categories that can be recognized, and decreasing the constraints on the images concerned (e.g. the number and occlusion of objects). However, these deep-learning systems will still share some of the weaknesses of their predecessors.

For instance, they—like the cat's-face recognizer—will have no understanding of 3D space, no knowledge of what a 'profile', or occlusion, actually is. Even vision programs designed for robots provide only an inkling of such matters.

The Mars Rover robots, such as *Opportunity* and *Curiosity* (landed in 2004 and 2012, respectively), rely on special knowledge-representation tricks: heuristics tailored for the 3D problems they're expected to face. They can't do pathfinding or object manipulation in the general case. Some robots simulate *animate* vision, wherein the body's own movements provide useful information (because they change the visual input systematically). But even they can't notice a possible pathway, or recognize that *this* unfamiliar thing could be picked up by their robot hand whereas *that* could not.

By the time this book is published, there may be some exceptions. But they too will have limits. For instance, they won't understand 'I can't pick that up', because they won't understand *can* and *cannot*. That's because the requisite modal logic probably still won't be available for their knowledge representation.

Sometimes, vision can ignore 3D space—when reading handwriting, for instance.

But even 2D computer vision is limited. Despite considerable research effort on *analogical*, or *iconic*, representations, AI can't reliably use diagrams in problem solving—as we do in geometrical reasoning, or in sketching abstract relationships on the back of an envelope. (Similarly, psychologists don't yet understand just how *we* do those things.)

In short, most human visual achievements surpass today's AI. Often, AI researchers aren't clear about what questions to ask. For instance, think about folding a slippery satin dress neatly. No robot can do this (although some can be instructed, step by step, how to fold an oblong terry towel). Or consider putting on a T-shirt: the head must go in first, and *not* via a sleeve—but *why*? Such topological problems hardly feature in AI.

None of this implies that human-level computer vision is impossible. But achieving it is much more difficult than most people believe.

So this is a special case of the fact noted in Chapter 1: that AI has taught us that human minds are hugely richer, and more subtle, than psychologists previously imagined. Indeed, that is *the* main lesson to be learned from AI.

The frame problem

Finding an appropriate knowledge representation, in whatever domain, is difficult partly because of the need to avoid the *frame problem*. (Beware: although this problem arises when using frames as a knowledge representation for concepts, the meanings of 'frame' here are different.)

As originally defined by McCarthy and Hayes, the frame problem involves assuming (during planning by robots) that an action will cause only *these* changes, whereas it may cause *those* too. More generally, the frame problem arises whenever implications tacitly assumed by human thinkers are ignored by the computer because they haven't been made explicit.

The classic case is the monkey and bananas problem, wherein the problem-solver (perhaps an AI planner for a robot) assumes that nothing relevant exists outside the frame (see Figure 1).

My own favourite example is: *If a man of 20 can pick 10 pounds of blackberries in an hour, and a woman of 18 can pick 8, how many will they gather if they go blackberrying together?* For sure, '*18*' isn't a plausible answer. It could be much more (because they're both showing off) or, more probably, much less. Just what kinds of knowledge are involved here? And could an AGI overcome what appear to be the plain arithmetical facts?

The frame problem arises because AI programs don't have a human's sense of *relevance* (see Chapter 3). It can be avoided if all possible consequences of every possible action are known. In some technical/scientific areas, that's so. In general, however, it isn't. That's a major reason why AI systems lack common sense.

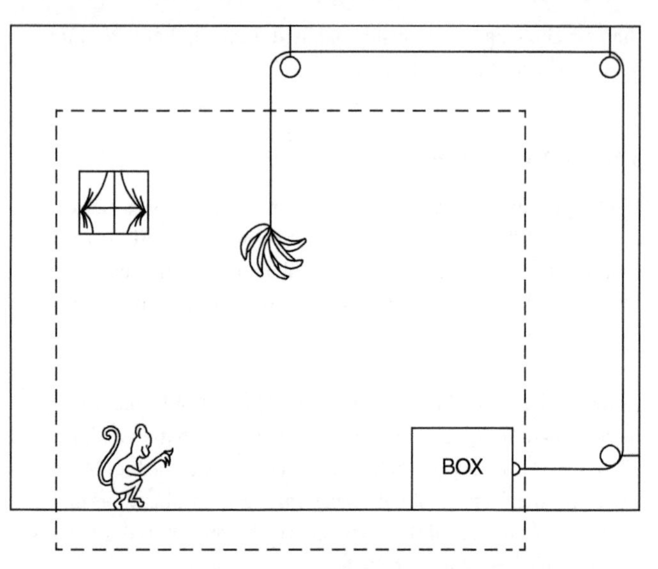

1. Monkey and bananas problem: how does the monkey get the bananas? (The usual approach to this problem assumes, though doesn't explicitly state, that the relevant 'world' is that shown inside the dotted-line frame. In other words, nothing exists outside this frame which causes significant changes in it on moving the box.)

In brief, the frame problem lurks all around us—and is a major obstacle in the quest for AGI.

Agents and distributed cognition

An AI *agent* is a self-contained ('autonomous') procedure, comparable sometimes to a knee-jerk reflex and sometimes to a mini-mind. Phone apps or spelling correctors could be called agents, but usually aren't—because agents normally *cooperate*. They use their highly limited intelligence in cooperation with—or anyway, alongside—others to produce results that they couldn't achieve alone. The interaction between agents is as important as the individuals themselves.

Some agent systems are organized by hierarchical control: top dogs and underdogs, so to speak. But many exemplify *distributed* cognition. This involves cooperation without hierarchical command structure (hence the prevarication, earlier, between 'in cooperation with' and 'alongside'). There's no central plan, no top-down influence, and no individual possessing *all* the relevant knowledge.

Naturally occurring examples of distributed cognition include ant trails, ship navigation, and human minds. Ant trails emerge from the behaviour of many individual ants, automatically dropping (and following) chemicals as they walk. Similarly, navigation and manoeuvring of ships results from the interlocking activities of many people: not even the captain has all the necessary knowledge, and some crew members have very little indeed. Even a single mind involves distributed cognition, for it integrates many cognitive, motivational, and emotional subsystems (see Chapters 4 and 6).

Artificial examples include neural networks (see Chapter 4); an anthropologist's computer model of ship navigation, and A-Life work on situated robotics, swarm intelligence, and swarm robotics (see Chapter 5); symbolic AI models of financial markets (the agents being banks, hedge funds, and large shareholders); and the LIDA model of consciousness (see Chapter 6).

Clearly, human-level AGI would involve distributed cognition.

Machine learning

Human-level AGI would include machine learning, too. However, this needn't be *human-like*. The field originated from psychologists' work on concept learning and reinforcement. However, it now depends on fearsomely mathematical techniques, because the knowledge representations used involve probability theory and statistics. (One might say that psychology has been left far behind. Certainly, some modern machine learning systems bear little or no similarity to what might plausibly be going on in human heads. However, the increasing use of *Bayesian* probability in this area of AI parallels recent theories in cognitive psychology and neuroscience.)

Today's machine learning is hugely lucrative. It's used for data mining—and, given supercomputers doing a million billion calculations per second, for processing Big Data (see Chapter 3).

Some machine learning uses neural networks. But much relies on symbolic AI, supplemented by powerful statistical algorithms. In fact, the statistics really do the work, the GOFAI merely guiding the worker to the workplace. Accordingly, some professionals regard machine learning as computer science and/or statistics—*not* AI. However, there's no clear boundary here.

Machine learning has three broad types: supervised, unsupervised, and reinforcement learning. (The distinctions originated in psychology, and different neurophysiological mechanisms may be involved; reinforcement learning, across species, involves dopamine.)

In *supervised* learning, the programmer 'trains' the system by defining a set of desired outcomes for a range of inputs (labelled examples and non-examples), and providing continual feedback about whether it has achieved them. The learning system generates hypotheses about the relevant features. Whenever it classifies incorrectly, it amends its hypothesis accordingly. *Specific* error messages are crucial (not merely feedback that it was mistaken).

In *unsupervised* learning, the user provides no desired outcomes or error messages. Learning is driven by the principle that co-occurring features engender expectations that they will co-occur in future. Unsupervised learning can be used to *discover* knowledge. The programmers needn't know what patterns/clusters exist in the data: the system finds them for itself.

Finally, *reinforcement* learning is driven by analogues of reward and punishment: feedback messages telling the system that what it just did was good or bad. Often, reinforcement isn't simply binary, but represented by numbers—like the scores in a video game. 'What it just did' may be a single decision (such as a move in a game), or a series of decisions (e.g. chess moves culminating in checkmate). In some video games, the numerical score is updated at every move. In highly complex situations, such as chess, success (or failure) is signalled only after many decisions, and some procedure for *credit assignment* identifies the decisions most likely to lead to success.

Symbolic machine learning in general assumes—what's not obviously true—that the knowledge representation for learning must involve some form of probability distribution. And many learning algorithms assume—what is usually false—that every variable in the data has the same probability distribution, and all are mutually independent. That's because this *i.i.d.* (independent and identically distributed) assumption underlies many mathematical theories of probability, on which the algorithms are based. The mathematicians adopted the i.i.d. assumption because it makes the mathematics simpler. Similarly, using i.i.d. in AI simplifies the search space, thus making problem solving easier.

Bayesian statistics, however, deals with *conditional* probabilities, where items/events are *not* independent. Here, probability depends on distributional evidence about the domain. Besides being more realistic, this form of knowledge representation allows probabilities to be changed if new evidence comes in. Bayesian techniques are becoming increasingly prominent in AI—and in psychology and neuroscience too. Theories of 'the Bayesian brain' (see Chapter 4) capitalize on the use of non-i.i.d. evidence to drive, and to fine-tune, unsupervised learning in perception and motor control.

Given various theories of probability, there are many different algorithms suitable for distinct types of learning and different data sets. For instance, Support Vector Machines—which accept the i.i.d. assumption—are widely used for supervised learning, especially if the user lacks specialized prior knowledge about the domain. 'Bag of Words' algorithms are useful when the *order* of features can be ignored (as in searches for words, not phrases). And if the i.i.d. assumption is dropped, Bayesian techniques ('Helmholtz machines') can learn from distributional evidence.

Most machine learning professionals use off-the-shelf statistical methods. The originators of those methods are highly prized by the industry: Facebook recently employed the creator of Support Vector Machines, and in 2013/14 Google hired several key instigators of *deep learning*.

Deep learning is a promising new advance based in multilayer networks (see Chapter 4), by which patterns in the input data are recognized at various hierarchical levels. In other words, deep learning *discovers* a multilevel knowledge representation—for instance, pixels to contrast detectors, to edge detectors, to shape detectors, to object parts, to objects.

One example is the cat's-face detector that emerged from Google's research on YouTube. Another, reported in *Nature* in 2015, is a reinforcement learner (the 'DQN' algorithm) that has learned to play the classic Atari 2600 2D games. Despite being given only pixels and game scores as input (and already knowing only the number of actions available for each game), this surpasses 75 per cent of humans on twenty-nine of the forty-nine games, and outperforms professional game testers on twenty-two.

It remains to be seen how far this achievement can be extended. Although DQN sometimes finds the optimal strategy, involving temporally ordered actions, it can't master games whose planning encompasses a longer period of time.

Future neuroscience may suggest improvements to this system. The current version is inspired by the Hubel–Wiesel vision receptors—cells in the visual cortex that respond only to movement, or only to lines of a particular orientation. (That's no big deal: the Hubel–Wiesel receptors inspired Pandemonium, too: see Chapter 1.) More unusually, this version of DQN is inspired also by the 'experience replay' happening in the hippocampus during sleep. Like the hippocampus, the DQN system stores a pool of past samples, or experiences, and reactivates them rapidly during learning. This feature is crucial: the designers reported 'a severe deterioration' in performance when it was disabled.

Generalist systems

The Atari game player caused excitement—and merited publication in *Nature*—partly because it seemed to be a step towards AGI. A single algorithm, using no handcrafted knowledge representation, learned a wide range of competences on a variety of tasks involving relatively high-dimensional sensory input. No previous program had done that.

Nor did the *AlphaGo* program, developed by the same team, which in 2016 beat the world's *Go* champion Lee Sedol. Nor *AlphaGo Zero*, which in 2017 surpassed *AlphaGo* despite having been fed no data about games of *Go* played by humans. For the record, in December 2017 *AlphaZero* also mastered chess: after only four hours of playing against itself, starting from random states but having been provided with the rules of the game, it beat the champion chess-program *Stockfish* by twenty-eight wins and seventy-two draws in one hundred games.

However (as remarked at the outset of this chapter), full AGI would do very much more. Difficult though it is to build a high-performing AI specialist, building an AI generalist is orders of magnitude harder. (Deep learning isn't the answer: its *aficionados* admit that 'new paradigms are needed' to combine it with complex reasoning—scholarly code for 'we haven't got a clue'.) That's why most AI researchers abandoned that early hope, turning instead to multifarious narrowly defined tasks—often with spectacular success.

AGI pioneers who retained their ambitious hopes included Newell and John Anderson. They originated SOAR and ACT-R respectively: systems begun in the early 1980s, and both still being developed (and used) some three decades later. However, they oversimplified the task, focusing on only a small subset of human competences.

In 1962, Newell's colleague Simon had considered the zig-zagging path of an ant on uneven ground. Every movement, he said, is a direct reaction to the situation perceived by the ant at that moment (this is the key idea of *situated* robotics: see Chapter 5). Ten years later, Newell and Simon's book *Human Problem Solving* described our intelligence as similar. According to their psychological theory, perception and motor action are supplemented by internal representations (IF–THEN rules, or 'productions') stored in memory, or newly built during problem solving.

'Human beings, viewed as behaving systems', they said, 'are quite simple'. But the emergent behavioural complexities are significant. For instance, they showed that a system of only fourteen IF–THEN rules can solve cryptarithmetic problems (e.g. map the letters to the digits 0 to 9 in this sum: DONALD + GERALD = ROBERT, where D = 5). Some rules deal with goal/sub-goal organization. Some direct attention (to a specific letter or column). Some recall previous steps (intermediate results). Some recognize false starts. And others backtrack to recover from them.

Cryptarithmetic, they argued, exemplifies the computational architecture of *all* intelligent behaviour—so this psychological approach suited a *generalist* AI. From 1980, Newell (with John Laird and Paul Rosenbloom) developed SOAR (Success Oriented Achievement Realized). This was intended as a model of cognition as a whole. Its reasoning integrated perception, attention, memory, association, inference, analogy, and learning. Ant-like (situated) responses were combined with internal deliberation. Indeed, deliberation often resulted in reflex responses, because a previously used sequence of sub-goals could be 'chunked' into *one* rule.

In fact, SOAR failed to model *all* aspects of cognition, and was later extended as people recognized some of the gaps. Today's version is used for many purposes, from medical diagnosis to factory scheduling.

Anderson's ACT-R (Adaptive Control of Thought) family are hybrid systems (see Chapter 4), developed by combining production systems and semantic networks. These programs, which recognize the statistical probabilities in the environment, model associative memory, pattern recognition, meaning, language, problem solving, learning, imagery, and (since 2005) perceptuo-motor control.

A key feature of ACT-R is the integration of procedural and declarative knowledge. Someone may *know that* a theorem of Euclid's is true, without *knowing how* to use it in a geometrical proof. ACT-R can learn how to apply a propositional truth, by constructing hundreds of new productions that control its use in many different circumstances. It learns which goals, sub-goals, and sub-sub-goals... are relevant in which conditions, and what results a particular action will have in various circumstances. In short, it learns by doing. And (like SOAR) it can chunk several rules that are often carried out sequentially into a single rule. This parallels the difference between how human experts and novices solve 'the same' problem: unthinkingly or painstakingly.

ACT-R has diverse applications. Its mathematics tutors offer personalized feedback, including relevant domain knowledge, and the goal/sub-goal structure of problem solving. Thanks to chunking, the grain size of their suggestions changes as the student's learning proceeds. Other applications concern NLP; human–computer interaction; human memory and attention; driving and flying; and visual web search.

SOAR and ACT were contemporaries of another early attempt at AGI: Douglas Lenat's CYC. This symbolic-AI system was launched in 1984, and is still under continuous development.

By 2015, CYC contained 62,000 'relationships' capable of linking the concepts in its database, and millions of links between those concepts. These include the semantic and factual associations stored in large semantic nets (see Chapter 3), and countless facts of naïve physics—the unformalized knowledge of physical phenomena (such as dropping and spilling) that all humans have. The system uses both monotonic and non-monotonic logics, and probabilities too, to reason about its data. (At present, all the concepts and links are hand-coded, but Bayesian learning is being added; this will enable CYC to learn from the Internet.)

It has been used by several US government agencies, including the Department of Defense (to monitor terrorist groups, for instance) and the National Institutes of Health, and by some major banks and insurance companies. A smaller version—*OpenCyc*—has been publicly released as a background source for a variety of applications, and a fuller abridgment (*ResearchCyc*) is available for AI workers. Although *OpenCyc* is regularly updated it contains only a small subset of CYC's database, and a small subset of inference rules. Eventually, the complete (or near-complete) system will be commercially available. However, that could fall into malicious hands—unless specific measures are taken to prevent this (see Chapter 7).

CYC was described by Lenat in *AI Magazine* (1986) as 'Using Common Sense Knowledge to Overcome Brittleness and Knowledge Acquisition Bottlenecks'. That is, it was specifically addressing McCarthy's prescient challenge. Today, it's the leader in modelling 'common-sense' reasoning, and also in 'understanding' the concepts it deals with (which even apparently impressive NLP programs cannot do: see Chapter 3).

Nevertheless, it has many weaknesses. For example, it doesn't cope well with metaphor (although the database includes many dead metaphors, of course). It ignores various aspects of naïve physics. Its NLP, although constantly improving, is very limited. And it doesn't yet include vision. In sum, despite its en-CYC-lopedic aims, it doesn't really encompass human knowledge.

The dream revitalized

Newell, Anderson, and Lenat beavered away in the background for thirty years. Recently, however, interest in AGI has revived markedly. An annual conference was started in 2008, and SOAR, ACT-R, and CYC are being joined by other supposedly generalist systems.

For instance, in 2010 the machine learning pioneer Tom Mitchell launched Carnegie Mellon's NELL (Never-Ending Language Learner). This 'common-sense' system builds its knowledge by trawling the Web non-stop (for seven years at the time of writing) and by accepting online corrections from the public. It can make simple inferences based on its (unlabelled) data: for instance, the athlete Joe Bloggs plays tennis, since he's on the Davis team. Starting with an ontology of 200 categories and relations (e.g. *master*, *is due to*), after five years it had enlarged the ontology and amassed ninety million candidate beliefs, each with its own confidence level.

The bad news is that NELL doesn't know, for example, that you can pull objects with a string, but not push them. Indeed, the putative common sense of *all* AGI systems is gravely limited. Claims that the notorious frame problem has been 'solved' are highly misleading.

NELL now has a sister program, NEIL: Never-Ending Image Learner. Some part-visual AGIs combine a logical-symbolic knowledge representation with analogical, or graphical, representations (a distinction made years ago by Aaron Sloman, but still not well understood).

In addition, Stanford Research Institute's CALO (Cognitive Assistant that Learns and Organizes) provided the spin-off *Siri* app (see Chapter 3), bought by Apple for $200 million in 2009. Comparable currently active projects include Stan Franklin's intriguing LIDA (discussed in Chapter 6) and Ben Goertzel's *OpenCog*, which learns its facts and concepts within a rich virtual world and also from other AGI systems. (LIDA is one of two generalist systems focused on *consciousness*; the other is CLARION.)

An even more recent AGI project, started in 2014, aims at developing 'A Computational Architecture for Moral Competence in Robots' (see Chapter 7). Besides the difficulties mentioned earlier, it will have to face the many problems that relate to morality.

A genuinely human-level system would do no less. No wonder, then, that AGI is proving so elusive.

Missing dimensions

Nearly all of today's generalist systems are focused on *cognition*. Anderson, for instance, aims to specify 'how all the subfields in cognitive psychology interconnect'. ('*All*' the subfields? Although he addresses motor control, he doesn't discuss touch or proprioception—which sometimes feature in robotics.) A truly general AI would cover *motivation* and *emotion* as well.

A few AI scientists have recognized this. Marvin Minsky and Sloman have both written insightfully about the computational architecture of whole minds, although neither has built a whole-mind model.

Sloman's MINDER model of anxiety is outlined in Chapter 3. His work (and Dietrich Dorner's psychological theory) has inspired Joscha Bach's *MicroPsi:* an AGI based on seven different 'motives', and using 'emotional' dispositions in planning and action selection. It has also influenced the LIDA system mentioned earlier (see Chapter 6).

But even these fall far short of true AGI. Minsky's prescient AI manifesto of 1956, 'Steps Toward Artificial Intelligence', identified obstacles as well as promises. Many of the former have yet to be overcome. As Chapter 3 should help to show, human-level AGI isn't within sight.

Chapter 3
Language, creativity, emotion

Some areas of AI seem especially challenging: language, creativity, and emotion. If AI can't model these, hopes of AGI are illusory.

In each case, more has been achieved than many people imagine. Nevertheless, significant difficulties remain. These quintessentially 'human' areas have been modelled only up to a point. (Whether AI systems could ever have *real* understanding, creativity, or emotion is discussed in Chapter 6. Here, our question is whether they can *appear* to possess them.)

Language

Countless AI applications use natural language processing (NLP). Most focus on the computer's 'understanding' of language that is presented to it, not on its own linguistic production. That's because NLP generation is even more difficult than NLP acceptance.

The difficulties concern both thematic content and grammatical form. For instance, we saw in Chapter 2 that familiar action sequences ('scripts') can be used as the seed of AI stories. But whether the background knowledge representation includes enough about human motivation to make the story interesting is another matter. A commercially available system that writes annual summaries describing a firm's changing financial position generates very boring 'stories'. Computer-generated novels and soap-opera plots do exist—but they won't win any prizes for subtlety. (AI translations/summaries of human-generated texts may be much richer, but that's thanks to the *human* authors.)

As for grammatical form, computer prose is sometimes grammatically incorrect and usually very clumsy. Anthony Davey's AI-generated narrative of a game of noughts and crosses (tic-tac-toe) can have clausal/sub-clausal structures that match the dynamics of the game in a nicely appropriate way. But the possibilities and strategies of noughts and crosses are fully understood. Describing the succession of thoughts, or actions, of the protagonists in most human stories in a similarly elegant fashion would be much more challenging.

Turning to AI's *acceptance* of language, some systems are boringly simple: they require only keyword recognition (think of the 'menus' in e-retailing), or the prediction of words listed in a dictionary (think of the automatic completion that happens when writing text messages). Others are significantly more sophisticated.

A few require speech recognition, either of single words, as in automated telephone shopping, or of continuous speech, as in real-time TV subtitling and telephone bugging. In the latter case, the aim may be to pick out specific words (such as *bomb* and *Jihad*) or, more interestingly, to capture the sense of the sentence as a whole. This is NLP with knobs on: the words themselves—spoken by many different voices, and with different local/foreign accents—must be distinguished first. (Word distinctions come for free in printed texts.) Deep learning (see Chapter 4) has enabled significant advances in speech processing.

Impressive examples of what looks like whole-sentence understanding include machine translation; data mining from large collections of natural-language texts; summarizing articles in newspapers and journals; and free-range question answering (increasingly employed in Google searches, and in the *Siri* app for the iPhone).

But can such systems really appreciate language? Can they cope with grammar, for instance?

In AI's early days, people assumed that language understanding requires syntactic parsing. Considerable effort went into writing programs to do that. The outstanding example—which brought AI to the attention of countless people who had previously never heard of it, or who had dismissed it as impossible—was Terry Winograd's SHRDLU, written at MIT in the early 1970s.

This program accepted instructions in English telling a robot to build structures made of coloured blocks, and worked out just how certain blocks should be moved to achieve the goal. It was hugely influential for many reasons, some of which applied to AI in general. Here, what's relevant is its unprecedented ability to assign detailed grammatical structure to complex sentences, such as: *How many eggs would you have been going to use in the cake if you hadn't learned your grandmother's recipe was wrong?* (Try it!)

For technological purposes, SHRDLU turned out to be a disappointment. The program contained many bugs, so could be used only by a handful of highly skilled researchers. Various other syntax crunchers were built at around that time, but they, too, weren't generalizable to real-world texts. In short, it soon appeared that the analysis of fancy syntax is too difficult for off-the-shelf systems.

Fancy syntax wasn't the only problem. In human language-use, *context* and *relevance* matter too. It wasn't obvious that they could ever be handled by AI.

Indeed, machine translation had been pronounced impossible by the US government's ALPAC Report in 1964 (the acronym denoted the Automatic Language Processing Advisory Committee). Besides predicting that not enough people would want to use it to make it commercially viable (although machine aids for human translators might admittedly be feasible), the report argued that computers would struggle with syntax, be defeated by context, and—above all—be blind to relevance.

That was a bombshell for machine translation (whose funding virtually dried up overnight), and for AI in general. It was widely interpreted as showing the futility of AI. The bestseller *Computers and Common Sense* had already claimed (in 1961) that AI was a waste of taxpayers' money. Now, it seemed that top governmental experts agreed. Two US universities that were about to open AI departments cancelled their plans accordingly.

Work in AI continued nevertheless, and when the syntax-savvy SHRDLU hit the scene a few years later it seemed to be a triumphant vindication of GOFAI. But doubts soon crept in. Accordingly, NLP turned increasingly to context rather than syntax.

A few researchers had taken semantic context seriously even in the early 1950s. Margaret Masterman's group in Cambridge, England, had approached machine translation (and information retrieval) by using a thesaurus rather than a dictionary. They saw syntax as 'that very superficial and highly redundant part of language that [people in a hurry], quite rightly, drop', and focused on word clusters rather than single words. Instead of attempting word-by-word translation, they searched the surrounding text for words of similar meaning. This (when it worked) enabled ambiguous words to be translated correctly. So *bank* could be rendered (in French) as *rive* or as *banque*, depending on whether the context contained words such as *water* or *money*, respectively.

That thesaurus-based contextual approach could be strengthened by also considering words that often co-occur despite having *dissimilar* meanings (like *fish* and *water*). And this, as time passed, is what happened. Besides distinguishing various types of lexical similarity—synonyms (*empty/vacant*), antonyms (*empty/full*), class membership (*fish/animal*) and inclusion (*animal/fish*), shared class level (*cod/salmon*), and part/whole (*fin/fish*)—today's machine translation also recognizes thematic co-occurrence (*fish/water*, *fish/bank*, *fish/chips*, etc.).

It's now clear that handling fancy syntax isn't necessary for summarizing, questioning, or translating a natural-language text. Today's NLP relies more on brawn (computational power) than on brain (grammatical analysis). Mathematics—specifically, statistics—has overtaken logic, and machine learning (including, but not restricted to, deep learning) has displaced syntactic analysis. These new approaches to NLP, ranging from written texts to speech recognition, are so efficient that a 95 per cent success rate is taken as the norm of acceptability for practical applications.

In modern-day NLP, powerful computers do statistical searches of huge collections ('corpora') of texts (for machine translation, these are paired translations done by humans) to find word patterns both commonplace and unexpected. They can learn the statistical likelihood of *fish/water*, or *fish/tadpole*, or *fish and chips/salt and vinegar*. And (as remarked in Chapter 2) NLP can now learn to construct 'word vectors' representing the probabilistic clouds of meaning that attend a given concept. In general, however, the focus is on words and phrases, not syntax. Grammar isn't ignored: labels such as ADJective and ADVerb may be assigned, either automatically or by hand, to some words in the texts being examined. But syntactic *analysis* is little used.

Even detailed *semantic* analysis isn't prominent. 'Compositional' semantics uses syntax in analysing the meaning of sentences; but it is found in research laboratories, not in large-scale applications. The 'common-sense' reasoner CYC has relatively full semantic representations of its concepts (words), and 'understands' them better accordingly (see Chapter 2). But that is still unusual.

Current machine translation can be astonishingly successful. Some systems are restricted to a small set of topics, but others are more open. Google Translate offers machine translation on unconstrained topics to over 200 million users every day. SYSTRAN is used daily by the European Union (for twenty-four languages) and NATO, and by Xerox and General Motors.

Many of these translations, including the EU documents, are near-perfect (because only a limited subset of words is used in the original texts). Many more are imperfect yet easily intelligible, because informed readers can ignore grammatical errors and inelegant word choices—as one does when listening to a non-native speaker. Some require minimal post-editing by humans. (With Japanese, significant pre-editing and post-editing may be needed. Japanese contains no segmented words, like the English past tense *vot-ed*, and phrase orderings are reversed. Machine-matching of languages from different language groups is usually difficult.)

In short, the results of machine translation are normally good enough for the human user to understand. Similarly, *monolingual* NLP programs that summarize journal papers can often show whether the paper merits reading in full. (*Perfect* translation is arguably impossible anyway. For example, requesting an apple in Japanese requires language reflecting the interlocutors' comparative social status, but no equivalent distinctions exist in English.)

The real-time translation available on AI applications such as Skype is less successful. That's because the system has to recognize speech, not written text (in which the individual words are clearly separated).

Two other prominent NLP applications are forms of information retrieval: *weighted search* (initiated by Masterman's group in 1976) and *data mining*. The Google search engine, for instance, searches for terms weighted by relevance—which is assessed statistically, not semantically (that is, *without* understanding).

Data mining can find word patterns unsuspected by human users. Long used for market research on products and brands, it's now being applied (often using deep learning) to 'Big Data': huge collections of texts (sometimes multilingual) or images, such as scientific reports, medical records, or entries on social media and the Internet.

Applications of Big Data mining include surveillance and counter-espionage, and the monitoring of public attitudes by governments, policy-makers, and social scientists. Such enquiries can compare the shifting opinions of distinct sub-groups: men/women, young/old, North/South, and so on. For instance, the UK think tank Demos (working with an NLP data-analytics team at the University of Sussex) has analysed many thousands of Twitter messages relating to misogyny, ethnic groups, and the police. Sudden bursts of tweeting after specific events ('twitcidents') can be searched to discover, for example, changes in public opinion on the police's reaction to a particular incident.

It remains to be seen whether Big Data NLP will reliably produce useful results. Often, data mining (using 'sentiment analysis') seeks to measure not only the level of public interest, but its evaluative tone. However, this isn't straightforward. For instance, a tweet containing an apparently derogatory racial epithet, and so machine-coded as 'negative' in sentiment, may not in fact be derogatory. A human judge, on reading it, may see the term as being used (in this case) as a positive marker of group identity, or as a neutral description (e.g. *The Paki shop on the corner*), not as insult or abuse. (The Demos research found that only a small proportion of tweets containing racial/ethnic terms are actually aggressive.)

In such cases, the human's judgement will rely on the context—for example, the other words within the tweet. It may be possible to adjust the machine's search criteria so that it makes fewer 'negative sentiment' ascriptions. Then again, it may not. Such judgements are often contentious. Even when they are agreed, it may be difficult to identify those aspects of the context which justify the human's interpretation.

That's just one example of the difficulty of pinning down *relevance* in computational (or even verbal) terms.

Two well-known NLP applications may seem, at first sight, to contradict that statement: Apple's *Siri* and IBM's WATSON.

Siri is a (rule-based) personal assistant, a talking 'chat-bot' that can quickly answer many different questions. It has access to everything on the Internet—including Google Maps, Wikipedia, the constantly updated *New York Times,* and lists of local services such as taxis and restaurants. It also calls on the powerful question answerer *WolframAlpha*, which can use logical reasoning to work out—not merely *find*—answers to a wide range of factual questions.

Siri accepts a spoken question from the user (to whose voice and dialect it gradually adapts), and answers it by using web-searching and conversational analysis. Conversational analysis studies how people organize the sequence of topics in a conversation, and how they arrange interactions such as explanation and agreement. This approach enables *Siri* to consider questions such as *What does the interlocutor want?* and *How should it answer?*, and—up to a point—to adapt to the individual user's interests and preferences.

In short, *Siri* appears to be sensitive not only to topical relevance, but to personal relevance as well. So it's superficially impressive. However, it's easily led into giving ridiculous answers—and if the user strays from the domain of facts, *Siri* is lost.

WATSON, too, is focused on facts. As an off-the-shelf resource (with 2,880 core processors) for handling Big Data, it's already used in some call centres, and it is being adapted for medical applications such as assessing cancer therapies. But it doesn't merely answer straightforward questions, as *Siri* does. It can also deal with the puzzles that arise in the general-knowledge game *Jeopardy!*.

In *Jeopardy!*, players aren't asked direct questions, but are given a clue and have to guess what the relevant question would be. For example, they are told 'On May 9, 1921, this "letter-perfect" airline opened its first passenger office in Amsterdam', and they should answer 'What is KLM?'

WATSON can meet that challenge, and many others. Unlike *Siri*, its *Jeopardy!*-playing version has no access to the Internet (although the medical version does), and no notion of the structure of conversations. Nor can it discover an answer by logical reasoning. Instead, it uses massively parallel statistical search over an enormous, but closed, database. This contains documents—countless reviews and reference books, plus the *New York Times*—providing facts about leprosy to Liszt, hydrogen to Hydra, and so on. When playing *Jeopardy!*, its search is guided by hundreds of specially crafted algorithms that reflect the probabilities inherent in the game. And it can learn from its human contestants' guesses.

In 2011, WATSON rivalled the Kasparov moment of its IBM cousin *Deep Blue* (see Chapter 2), by apparently beating the two top human champions. ('Apparently', because the computer reacts instantaneously whereas humans need some reaction time before pressing the buzzer.) But, like *Deep Blue*, it doesn't always win.

On one occasion it lost because, although it correctly focused on a particular athlete's *leg*, it didn't realize that the crucial fact in its stored data was that this person had a leg *missing*. That mistake won't reoccur, because WATSON's programmers have now flagged the importance of the word 'missing'. But others will. Even in mundane fact-seeking contexts, people often rely on relevance judgements that are beyond WATSON. For example, one clue required the identity of two of Jesus' disciples whose names are both top-ten baby names, and end in the same letter. The answer was 'Matthew and Andrew'—which WATSON got immediately. The human champion got that answer too. But his first idea had been 'James and Judas'. He rejected that, he recalled, only because 'I don't think Judas is a popular baby name, for some reason'. WATSON couldn't have done that.

Human judgements of relevance are often much less obvious than that one, and much too subtle for today's NLP. Indeed, relevance is a linguistic/conceptual version of the unforgiving 'frame problem' in robotics (see Chapter 2). Many people would argue that it will never be wholly mastered by a non-human system. Whether that's due only to the massive complexity involved, or to the fact that relevance is rooted in our specifically human form of life, is discussed in Chapter 6.

Creativity

Creativity—the ability to produce ideas or artefacts that are new, surprising, and valuable—is the acme of human intelligence, and necessary for human-level AGI. But it's widely seen as mysterious. It's not obvious how novel ideas could arise in *people,* never mind computers.

Even *recognizing* it isn't straightforward: people often disagree about whether an idea is creative. Some disagreements turn on whether, and in what sense, it's actually new. An idea may be new only to the individual involved, or new also to the whole of human history (exemplifying 'individual' and 'historical' creativity respectively). In either case, it may be *more* or *less* similar to preceding ideas, leaving room for further disagreements. Other disputes turn on valuation (which involves functional, and sometimes phenomenal, consciousness: see Chapter 6). An idea may be valued by one social group, but not others. (Think of the scorn directed by youngsters today to anyone who prizes their DVDs of Abba.)

It's commonly assumed that AI could have nothing interesting to say about creativity. But AI technology has generated many ideas that are historically new, surprising, and valuable. These arise, for instance, in designing engines, pharmaceuticals, and various types of computer art.

Moreover, AI concepts help to explain *human* creativity. They enable us to distinguish three types: combinational, exploratory, and transformational. These involve different psychological mechanisms, eliciting different sorts of surprise.

In *combinational* creativity, familiar ideas are combined in unfamiliar ways. Examples include visual collage, poetic imagery, and scientific analogies (the heart as a pump, the atom as a solar system). The new combination provides a statistical surprise: it was improbable, like an outsider winning the Derby. But it's intelligible, so valuable. *Just how valuable* depends on judgements of relevance, discussed earlier.

Exploratory creativity is less idiosyncratic, for it exploits some culturally valued way of thinking (e.g. styles of painting or music, or sub-areas of chemistry or mathematics). The stylistic rules are used (largely unconsciously) to produce the new idea—much as English grammar generates new sentences. The artist/scientist may explore the style's potential in an unquestioning way. Or they may deliberately push and test it, discovering what it can and cannot generate. It may even be tweaked, by slightly altering (e.g. weakening/strengthening) a rule. The novel structure, despite its novelty, will be recognized as lying within a familiar stylistic family.

Transformational creativity is a successor of exploratory creativity, usually triggered by frustration at the limits of the existing style. Here, one or more stylistic constraints are radically altered (dropped, negated, complemented, substituted, added...), so that novel structures are generated which *could not* have been generated before. These new ideas are deeply surprising, because they're seemingly *impossible*. They're often initially unintelligible, for they can't be fully understood in terms of the previously accepted way of thinking. However, they must be intelligibly close to the previous way of thinking if they are to be accepted. (Sometimes, this recognition takes many years.)

All three types of creativity occur in AI—often, with results attributed by observers to humans (in effect, passing the Turing Test: see Chapter 6). But they aren't found in the proportions one might expect.

In particular, there are very few combinational systems. One might think it's easy to model combinational creativity. After all, nothing could be simpler than making a computer produce unfamiliar associations of already stored ideas. The results will often be historically novel, and (statistically) surprising. But if they're also to be valuable, they must be mutually relevant. That's not straightforward, as we've seen. The joke-generating programs mentioned in Chapter 2 use joke templates to help provide relevance. Similarly, symbolic AI's *case-based reasoning* constructs analogies thanks to pre-coded structural similarities. So, their 'combinational' creativity has a strong admixture of exploratory creativity as well.

Conversely, one might expect that AI could never model transformational creativity. This expectation, also, is mistaken. Certainly, any program can do only what it's potentially capable of doing. But evolutionary programs can transform themselves (see Chapter 5). They can even evaluate their newly transformed ideas—but only *if* the programmer has provided clear criteria for selection. Such programs are routinely used for novelty-seeking AI applications—such as designing new scientific instruments or drugs.

This isn't a magic road to AGI, however. Valuable results are rarely guaranteed. Some evolutionary programs (in maths or science) can reliably find the optimal solution, but many problems can't be defined by optimization. Transformational creativity is risky, because previously accepted rules are broken. Any new structures must be evaluated, or chaos ensues. But current AI's fitness functions are defined by humans: the programs can't adapt/evolve them independently.

Exploratory creativity is the type best suited to AI. There are countless examples. Some exploratory AI novelties in engineering (including one generated by a program from CYC's designer: see Chapter 2) have been awarded patents. Although a patented idea isn't 'obvious to a person skilled in the art', it may unexpectedly lie within the potential of the style being explored. A few AI explorations are indistinguishable from outstanding human achievements—such as the composition, by David Cope's programs, of music in the style of Chopin or Bach. (How many *humans* can do that?)

However, even exploratory AI depends crucially on human judgement. For someone must recognize—and clearly state—the stylistic rules concerned. That's usually difficult. A world expert on Frank Lloyd Wright's Prairie Houses abandoned his attempt to describe their architectural style, declaring it 'occult'. Later, a computable 'shape-grammar' generated indefinitely many Prairie House designs, including the forty-odd originals—and *no* implausibilities. But the human analyst was ultimately responsible for the system's success. Only if an AGI could analyse styles (in art or science) *for itself* would its creative explorations be 'all its own work'. Despite some recent—very limited—examples of art styles being recognized by deep learning (see Chapters 2 and 4), that's a tall order.

AI has enabled human artists to develop a new art form: computer-generated (CG) art. This concerns architecture, graphics, music, choreography, and—less successfully (given NLP's difficulties with syntax and relevance)—literature. In CG art, the computer isn't a mere tool, comparable to a new paintbrush, helping the artist to do things they might have done anyway. Rather, the work couldn't have been done, or perhaps even imagined, without it.

CG art exemplifies all three types of creativity. For the reasons given earlier, hardly any CG art is combinational. (Simon Colton's *The Painting Fool* has produced visual collages related to war—but it was specifically instructed to search for images associated with 'war', which were readily available in its database.) Most is exploratory or transformational.

Sometimes, the computer generates the artwork entirely independently, by executing the program written by the artist. So Harold Cohen's AARON produces line drawings and coloured images unaided (sometimes generating colours so daringly beautiful that Cohen says it's a better colourist than he is himself).

In interactive art, by contrast, the form of the final artwork depends partly on input from the audience—who may or may not have deliberate control over what happens. Some interactive artists see the audience as fellow-creators, others as mere causal factors who unknowingly affect the artwork in various ways (and some, such as Ernest Edmonds, take both approaches). In evolutionary art, exemplified by William Latham and Jon McCormack, the results are continually generated/transformed by the computer—but the *selection* is usually done by artist or audience.

In short, AI creativity has many applications. It can sometimes match, or even exceed, human standards in some small corner of science or art. But matching human creativity *in the general case* is quite another matter. AGI is as far away as ever.

AI and emotion

Emotion, like creativity, is something usually seen as utterly alien to AI. Besides the intuitive implausibility, the fact that moods and emotions depend on neuromodulators diffusing in the brain seems to rule out AI models of affect.

For many years, AI scientists themselves appeared to agree. With a few early exceptions in the 1960s and 1970s—namely Herbert Simon, who saw emotion as involved in cognitive control, and Kenneth Colby, who built interesting, although grossly overambitious, models of neurosis and paranoia—they ignored emotion.

Today, things are different. Neuromodulation has been simulated (in GasNets: see Chapter 4). Moreover, many AI research groups are now addressing emotion. Most (not quite all) of this research is theoretically shallow. And most is potentially lucrative, being aimed at developing 'computer companions'.

These are AI systems—some screen-based, some ambulatory robots—designed to interact with people in ways that (besides being practically helpful) are affectively comfortable, even satisfying, for the user. Most are aimed at the elderly and/or disabled, including people with incipient dementia. Some are targeted on babies or infants. Others are interactive 'adult toys'. In short: computer carers, robot nannies, and sexual playmates.

The human–computer interactions concerned include: offering reminders about shopping, medication, and family visits; talking about, and helping to compile, a continuing personal journal; scheduling and discussing TV programs, including the daily news; making/fetching food and drink; monitoring vital signs (and babies' crying); and speaking and moving in sexually stimulating ways.

Many of these tasks will involve emotion on the person's part. As for the AI companion, this may be able to recognize emotions in the human user and/or it may respond in apparently emotional ways. For instance, sadness in the user—caused, perhaps, by mention of a bereavement—might elicit some show of sympathy from the machine.

AI systems can already recognize human emotions in various ways. Some are physiological: monitoring the person's breathing rate and galvanic skin response. Some are verbal: noting the speaker's speed and intonation, as well as their vocabulary. And some are visual: analysing their facial expressions. At present, all these methods are relatively crude. The user's emotions are both easily missed and easily misinterpreted.

Emotional performance on the computer companion's part is usually verbal. It's based in vocabulary (and intonation, if the system generates speech). But, much as the system watches out for familiar keywords from the user, so it responds in highly stereotyped ways. Occasionally, it may quote a human-authored remark or poem associated with something the user has said—perhaps in the diary. But the difficulties of NLP imply that computer-generated text is unlikely to be subtly appropriate. It may not even be acceptable: the user may be irritated and frustrated by a companion incapable of offering even the *appearance* of true companionship. Similarly, a purring robot cat may annoy the user, instead of communicating comfortably relaxed contentment.

Then again, it may not: *Paro*, a cuddly interactive 'baby seal' with charming black eyes and luxurious eyelashes, appears to be beneficial for many elderly people and/or people with dementia. (Future versions will monitor vital signs, alerting the person's human carers accordingly.)

Some AI companions can use their own facial expressions, and eye gaze, to respond in seemingly emotional ways. A few robots possess flexible 'skin', overlying a simulacrum of human facial musculature, whose configuration can suggest (to the human observer) up to a dozen basic emotions. The screen-based systems often show the face of a virtual character, whose expressions change according to the emotions it (he/she?) is supposedly undergoing. However, all these things risk falling into the so-called 'uncanny valley': people typically feel uncomfortable, or even deeply disturbed, when encountering creatures that are very similar to human beings *but not quite similar enough*. Robots, or screen avatars, with not-quite-human faces may therefore be experienced as threatening.

Whether it's ethical to offer such quasi-companionship to emotionally needy people is questionable (see Chapter 7). Certainly, some human–computer interactive systems (e.g. *Paro*) appear to provide pleasure, and even lasting contentment, to people whose lives seem otherwise empty. But is that enough?

There's scant theoretical depth to 'companion' models. The emotional aspects of AI companions are being developed for commercial purposes. There's no attempt to make them use emotions in solving their own problems, nor to illuminate the role that emotions play in the functioning of the mind as a whole. It's as though emotions are seen by these AI researchers as optional extras: to be disregarded unless, in some messily human context, they're unavoidable.

That dismissive attitude was widespread in AI until relatively recently. Even Rosalind Picard's work on 'affective computing', which brought emotions in from the cold in the late 1990s, didn't analyse them in depth.

One reason why AI ignored emotion (and Simon's insightful remarks about it) for so long is that most psychologists and philosophers did so too. In other words, they didn't think of *intelligence* as something that requires emotion. To the contrary, affect was assumed to disrupt problem solving and rationality. The idea that emotion can help one to decide what to do, and how best to do it, wasn't fashionable.

It eventually became more prominent, thanks partly to developments in clinical psychology and neuroscience. But its entry into AI was due also to two AI scientists, Marvin Minsky and Aaron Sloman, who had long considered *the mind as a whole,* rather than confining themselves—like most of their colleagues—to one tiny corner of mentality.

For instance, Sloman's ongoing *CogAff* project focuses on emotion's role in the computational architecture of the mind. *CogAff* has influenced the LIDA model of consciousness, released in 2011 and still being extended (see Chapter 6). It has also inspired the MINDER program, initiated by Sloman's group in the late 1990s.

MINDER simulates (the functional aspects of) the anxiety that arises within a nursemaid, left to look after several babies single-handedly. She/it has only a few tasks: to feed them, to try to prevent them from falling into ditches, and to take them to a first-aid station if they do. And she has only a few motives (goals): feeding a baby; putting a baby behind a protective fence, if one already exists; moving a baby out of a ditch for first aid; patrolling the ditch; building a fence; moving a baby to a safe distance from the ditch; and, if no other motive is currently activated, wandering around the nursery.

So she's hugely simpler than a real nursemaid (although more complex than a typical planning program, which has only one final goal). Nevertheless, she's prone to emotional perturbations comparable to various types of anxiety.

The simulated nursemaid has to respond appropriately to visual signals from her environment. Some of these trigger (or influence) goals that are more urgent than others: a baby crawling towards the ditch needs her attention sooner than a merely hungry baby, and one who's about to topple into the ditch needs it sooner still. But even those goals that can be put on hold may have to be dealt with eventually, and their degree of urgency may increase with time. So, a starving baby can be put back into its cot if another baby is near the ditch; but the baby who has waited longest for food should be nurtured before those fed more recently.

In short, the nursemaid's tasks can sometimes be interrupted, and either abandoned or put on hold. MINDER must decide just what the current priorities are. Such decisions must be taken throughout the session, and can result in repeated changes of behaviour. Virtually no task can be completed without interruption, because the environment (the babies) puts so many conflicting, and ever-changing, demands upon the system. As with a real nursemaid, the anxieties increase, and the performance degrades, with an increase in the number of babies—each of which is an unpredictable autonomous agent. Nevertheless, the anxiety is useful, for it enables the nursemaid to nurture the babies successfully. Successfully, but not *smoothly:* calm and anxiety are poles apart.

MINDER indicates some ways in which emotions can control behaviour, scheduling competing motives intelligently. A human nursemaid, no doubt, will experience various types of anxiety as her situation changes. But the point, here, is that emotions aren't merely *feelings*. They involve functional, as well as phenomenal, consciousness (see Chapter 6). Specifically, they are computational mechanisms that enable us to schedule competing motives—and without which we couldn't function. (So the emotionless Mr Spock of *Star Trek* is an evolutionary impossibility.)

If we are ever to achieve AGI, emotions such as anxiety will have to be included—and *used*.

Chapter 4
Artificial neural networks

Artificial neural networks (ANNs) are made up of many interconnected units, each one capable of computing only one thing. Described in this way, they may sound boring. But they can seem almost magical. They've certainly bewitched the journalists. Frank Rosenblatt's 'perceptrons', photoelectric machines that learned to recognize letters without being explicitly taught, were puffed enthusiastically in the 1960s newspapers. ANNs made an especially noisy splash in the mid-1980s, and are still regularly hailed in the media. The most recent ANN-related hype concerns deep learning.

ANNs have myriad applications, from playing the stock market and monitoring currency fluctuations to recognizing speech or faces. But it's *the way they work* that is so intriguing.

A tiny handful are run on specifically parallel hardware—or even on a hardware/wetware mix, combining real neurons with silicon circuits. Usually, however, the network is simulated by a von Neumann machine. That is, ANNs are parallel-processing virtual machines implemented on classical computers (see Chapter 1).

They are intriguing partly because they are very different from the virtual machines of symbolic AI. Sequential instructions are replaced by massive parallelism, top-down control by bottom-up processing, and logic by probability. And the dynamical, continuously changing aspect of ANNs contrasts starkly with symbolic programs.

Moreover, many networks have the uncanny property of self-organization from a random start. (The 1960s perceptrons had this too: hence their high press profile.) The system starts with a random architecture (random weights and connections), and gradually adapts itself to perform the task required.

Neural networks have many strengths, and have added significant computational capabilities to AI. Nevertheless, they also have weaknesses. So they can't deliver the truly *general* AI envisaged in Chapter 2. For instance, although some ANNs can do approximate inference, or reasoning, they can't represent precision as well as symbolic AI can. (*Q: What's 2 + 2? A: Very probably 4.* Really?) Hierarchy, too, is more difficult to model in ANNs. Some (*recurrent*) nets can use interacting networks to represent hierarchy—but only to a limited degree.

Thanks to the current enthusiasm for deep learning, networks of networks are less rare now than they used to be. However, they are still relatively simple. The human brain must comprise countless networks, on many different levels, interacting in highly complex ways. In short, AGI is still far distant.

The wider implications of ANNs

ANNs are a triumph of AI considered as computer science. But their theoretical implications go much further. Because of some general similarities to human concepts and memory, ANNs are of interest to neuroscientists, psychologists, and philosophers.

The neuroscientific interest isn't new. Indeed, the pioneering perceptrons were intended by Rosenblatt not as a source of practically useful gizmos, but as *a neuropsychological theory*.

Today's networks—despite their many differences from the brain—are important in computational neuroscience.

Psychologists, too, are interested in ANNs—and the philosophers haven't been far behind. For instance, one mid-1980s example caused a furore well outside the ranks of professional AI. This network apparently learned to use the past tense much as children do, starting by making no mistakes but then over-regularizing—so that *go/went* gives way to *go/goed*—before achieving correct usage for both regular and irregular verbs. That was possible because the input that was provided to it mirrored the changing probabilities of the words typically heard by a child: the network *wasn't* applying innate grammatical rules.

This was important because most psychologists (and many philosophers) at the time had accepted Noam Chomsky's claims that children *must* rely on inborn linguistic rules in order to learn grammar, and that infantile over-regularizations were irrefutable evidence of those rules being put to work. The past-tense network proved that neither of these claims is true. (It didn't prove, of course, that children don't have innate rules: merely that they don't *need* to have them.)

Another widely interesting example, which was originally inspired by developmental psychology, is research on 'representational trajectories'. Here (as also in deep learning), input data that are initially confusing are recoded on successive levels, so that less obvious regularities are captured in addition to the prominent ones. This relates not only to child development, but also to psychological and philosophical debates about inductive learning. For it shows that prior expectations (computational structure) are needed in order to learn patterns in the input data, and that there are unavoidable constraints on the order in which different patterns are learned.

In short, this AI methodology is theoretically interesting in many ways, as well as being hugely important commercially.

Parallel distributed processing

One category of ANNs in particular attracts huge attention: those doing PDP. Indeed, when people refer to 'neural networks' or 'connectionism' (a term less often used today), they usually mean PDP.

Because of the way they work, PDP networks share four major strengths. These relate to both technological applications and theoretical psychology (and also to the philosophy of mind).

The first is their ability to learn patterns, and associations between patterns, by being shown examples instead of being explicitly programmed.

The second is their tolerance of 'messy' evidence. They can do *constraint satisfaction*, making sense of partially conflicting evidence. They don't demand rigorous definitions, expressed as lists of necessary and sufficient conditions. Rather, they deal with overlapping sets of family resemblances—a feature of human concepts, too.

Yet another strength is their ability to recognize incomplete and/or partly damaged patterns. That is, they have *content-addressable* memory. So do people: think of identifying a melody from the first few notes, or played with many mistakes.

And fourth, they are robust. A PDP network with some nodes missing doesn't spout nonsense, or halt. It shows *graceful degradation*, in which performance worsens gradually as the damage increases. So they aren't brittle, as symbolic programs are.

These benefits result from the D in PDP. Not all ANNs involve distributed processing. In *localist* networks (such as *WordNet*: see Chapter 2), concepts are represented by single nodes. In *distributed* networks, a concept is stored across (distributed over) the whole system. Localist and distributed processing are sometimes combined, but that's uncommon. Purely localist networks are uncommon too, because they lack the major strengths of PDP.

One could say that distributed networks are localist *at base*, for each unit corresponds to a single microfeature—for example, a tiny patch of colour, at a particular place in the visual field. But these are defined at a much lower level than are concepts: PDP involves 'sub-symbolic' computation. Moreover, each unit can be part of many different overall patterns, so contributes to many different 'meanings'.

There are many types of PDP systems. All are made of three or more layers of interconnected units, each unit capable of computing only one simple thing. But the units differ.

A unit in the input layer fires whenever its microfeature is presented to the network. An output unit fires when it's triggered by the units connected to it, and its activity is communicated to the human user. The hidden units, in the middle layer(s), have no direct contact with the outside world. Some are *deterministic*: they fire, or not, depending only on the influences from their connections. Others are *stochastic*: whether they fire depends partly on some probability distribution.

The connections differ, too. Some are *feedforward*, passing signals from a lower layer to a higher one. Some send *feedback* signals in the opposite direction. Some are *lateral*, linking units within the same layer. And some, as we'll see, are both *feedforward* and *feedback*. Like brain synapses, connections are either excitatory or inhibitory. And they vary in strength, or *weight*. Weights are expressed as numbers between +1 and –1. The higher the weight of an excitatory (or inhibitory) link, the higher (or lower) the probability that the unit receiving the signal will fire.

PDP involves *distributed* representation, for each concept is represented by the state of the entire network. This may seem puzzling, even paradoxical. It's certainly very different from how representations are defined in symbolic AI.

People interested only in technological/commercial applications don't care about that. If they're satisfied that certain obvious questions—such as how a single network can store several different concepts, or patterns—aren't problematic in practice, they're happy to leave it at that.

People concerned with the psychological and philosophical implications of AI ask that 'obvious question', too. The answer is that the possible overall states of a PDP network are so multifarious that only a few will involve simultaneous activation in *this* or *that* scattering of units. An activated unit will spread activation only to *some* other units. However, those 'other units' vary: any given unit can contribute to many different patterns of activation. (In general, 'sparse' representations, with many unactivated units, are more efficient.) The system will saturate eventually: theoretical research on associative memories asks how many patterns can, in principle, be stored by networks of a certain size.

But those engaged with the psychological and philosophical aspects aren't happy to leave it at that. They are interested also in the concept of *representation* itself, and in debates about whether human minds/brains actually do contain internal representations. PDP devotees argue, for example, that this approach refutes the Physical Symbol System hypothesis, which originated in symbolic AI and rapidly spread into the philosophy of mind (see Chapter 6).

Learning in neural networks

Most ANNs can learn. This involves making adaptive changes in the weights, and sometimes also in the connections. Usually, the network's anatomy—the number of units, and the links between them—is fixed. If so, learning alters only the weights. But sometimes, learning—or evolution (see Chapter 5)—can add new links and prune old ones. *Constructive* networks take this to the extreme: starting with no hidden units at all, they add them as learning proceeds.

PDP networks can learn in many different ways—and exemplify all the types distinguished in Chapter 2: supervised, unsupervised, and reinforcement learning.

In supervised learning, for instance, they come to recognize a class by being shown various examples of it—none of which needs to possess *every* 'typical' feature. (The input data may be visual images, verbal descriptions, sets of numbers....) When an example is presented, some input units respond to 'their' microfeatures, and activations spread until the network settles down. The resulting state of the output units is then compared with the desired output (identified by the human user), and further weight changes are instigated (perhaps by *backprop*) so as to make those errors less probable. After many examples, differing slightly from each other, the network will have developed an activation pattern that corresponds to the typical case, or 'prototype', even if no such case has actually been encountered. (If a damaged example is now presented, stimulating many fewer of the relevant input units, this pattern will be completed automatically.)

Most ANN learning is based on the *fire together, wire together* rule, stated in the 1940s by the neuropsychologist Donald Hebb. Hebbian learning strengthens often-used connections. When two linked units are activated simultaneously, the weights are adjusted to make this more likely in future.

Hebb expressed the *ft/wt* rule in two ways, which were neither precise nor equivalent. Today's AI researchers define it in many different ways, based perhaps on differential equations drawn from physics, or on Bayesian probability theory. They use theoretical analysis to compare, and improve, the various versions. So, PDP research can be fiendishly mathematical.

Given that a PDP network is using some Hebbian learning rule to adapt its weights, when does it stop? The answer isn't *When it has achieved perfection (all inconsistencies eliminated),* but *When it has achieved maximum coherence.*

An 'inconsistency' occurs, for instance, when two microfeatures that aren't usually present together are simultaneously signalled by the relevant units. Many symbolic AI programs can do constraint satisfaction, approaching the solution by eliminating contradictions between evidence on the way. But they don't tolerate inconsistency as part of the solution. PDP systems are different. As the PDP strengths listed earlier show, they can perform successfully even if discrepancies persist. Their 'solution' is the overall state of the network when inconsistencies have been minimized, not abolished.

One way of achieving that is to borrow the idea of *equilibrium* from thermodynamics. Energy levels in physics are expressed numerically, as are the weights in PDP. If the learning rule parallels the physical laws (and if the hidden units are stochastic), the same statistical Boltzmann equations can describe the changes in both cases.

PDP can even borrow the method used to cool metals rapidly but evenly. Annealing starts at a high temperature and cools down gradually. PDP researchers sometimes use *simulated annealing,* wherein the weight changes in the first few cycles of equilibration are much larger than those in later cycles. This enables the network to escape from situations ('local minima') where overall consistency has been achieved relative to what went before, but even greater consistency (and a more stable equilibrium) could be reached if the system were disturbed. Compare shaking a bag of marbles, to dislodge any marbles resting on an internal ridge: one should start by shaking forcefully, but end by shaking gently.

A faster, and more widely used, way of achieving maximal consistency is to employ backprop. But whichever of the many learning rules is employed, the state *of the whole network* (and especially of the output units), at equilibrium, is taken to be the representation of the concept involved.

Backprop and brains—and deep learning

PDP enthusiasts argue that their networks are more biologically realistic than symbolic AI. It's true that PDP is inspired by brains, and that some neuroscientists use it to model neural functioning. However, ANNs differ significantly from what lies inside our heads.

One difference between (most) ANNs and brains is back-propagation, or backprop. This is a learning rule—or rather, a general class of learning rules—that's frequently used in PDP. Anticipated by Paul Werbos in 1974, it was defined more useably by Geoffrey Hinton in the early 1980s. It solves the problem of *credit assignment*.

This problem arises across all types of AI, especially when the system is continually changing. Given a complex AI system that's successful, *just which* parts of it are most responsible for the success? In evolutionary AI, credit is often assigned by the 'bucket-brigade' algorithm (see Chapter 5). In PDP systems with deterministic (not stochastic) units, credit is typically assigned by backprop.

The backprop algorithm traces responsibility back from the output layer into the hidden layers, identifying the individual units that need to be adapted. (The weights are updated to minimize prediction errors.) The algorithm needs to know the precise state of the output layer when the network is giving the right answer. (So backprop is supervised learning.) Unit-by-unit comparisons are made between this exemplary output and the output actually obtained from the network. Any difference between an output unit's activity in the two cases counts as an error.

The algorithm assumes that error in an output unit is due to error(s) in the units connected to it. Working backwards through the system, it attributes a specific amount of error to each unit in the first hidden layer, depending on the connection weight between it and the output unit. Blame is shared between all the hidden units connected to the mistaken output unit. (If a hidden unit is linked to several output units, its mini-blames are summed.) Proportional weight changes are then made to the connections between the hidden layer and the *preceding* layer.

That layer may be another (and another...) stratum of hidden units. But ultimately it will be the input layer, and the weight changes will stop. This process is iterated until the discrepancies at the output layer are minimized.

For many years, backprop was used only on networks with one hidden layer. Multilayer networks were rare: they are difficult to analyse, and even to experiment with. Recently, however, they have caused huge excitement—and some irresponsible hype—by the advent of deep learning. Here, a system learns structure reaching deep into a domain, as opposed to mere superficial patterns. In other words, it discovers a multilevel knowledge representation, not a single-level one.

Deep learning is exciting because it promises to enable ANNs, at last, to deal with hierarchy. Since the early 1980s, connectionists such as Hinton and Jeff Elman had struggled to represent hierarchy—by combining local/distributed representation, or by defining recurrent nets. (Recurrent nets, in effect, perform as a *sequence* of discrete steps. Recent versions, using deep learning, can sometimes predict the next word in a sentence, or even the next 'thought' in a paragraph.) But they had limited success (and ANNs still aren't suitable for representing precisely defined hierarchies or deductive reasoning).

Deep learning, too, was initiated in the 1980s (by Jurgen Schmidhuber). But the field exploded much more recently, when Hinton provided an efficient method enabling multilayer networks to discover relationships on many levels. His deep-learning systems are made of 'restricted' Boltzmann machines (no lateral connections) on half a dozen layers. First, the layers do unsupervised learning. They are trained one by one, using simulated annealing. The output of one layer is used as the input to the next. When the final layer has stabilized, the whole system is fine-tuned by backprop, reaching down through all the levels to assign credit appropriately.

This approach to learning is interesting to cognitive neuroscientists, as well as to AI technologists. That is because it specifies 'generative models' that learn to predict the (likeliest) causes of inputs to the network—thus providing a model of what Helmholtz in 1867 called 'perception as unconscious inference'. That is, perception is not a matter of passively receiving input from the sense organs. It involves active interpretation, and even anticipatory prediction, of that input. In brief, the eye/brain is not a camera.

Hinton joined Google in 2013, so backprop will be very busy. Google is already using deep learning in many applications, including speech recognition and image processing. Moreover, in 2014 it bought *DeepMind*, whose DQN algorithm mastered the classic Atari games by combining deep learning with reinforcement learning and whose *AlphaGo* program beat the world champion in 2016 (see Chapter 2). IBM also favours deep learning: WATSON uses it, and is being borrowed for many specialist apps (see Chapter 3).

However, if deep learning is undeniably useful that doesn't mean that it's well understood. Many different multilayer learning rules are being explored experimentally, but theoretical analysis is confused.

Among the countless unanswered questions is whether there is a depth sufficient for near-human performance. (The cat's-face unit mentioned in Chapter 2 resulted from a nine-layer system.) The human visual system, for instance, has seven anatomical levels: but how many are added by computations in the cerebral cortex? Since ANNs are inspired by brains (a point constantly stressed in deep-learning hype), that question is natural. But it's not quite so pertinent as it may seem.

Backprop is a computational triumph. But it's highly non-biological. No cat's-face 'grandmother cell' in the brain (see Chapter 2) could result from processes just like those in deep learning. Real synapses are purely feedforward: they don't transmit in both directions. Brains contain feedback *connections* in various directions, but each one is strictly one-way. That's just one of the many differences between real and artificial neural networks. (Another is that brain networks aren't organized as strict hierarchies—even though the visual system is often described that way.)

The fact that brains contain both forward and backward connections is crucial to *predictive coding* models of sensorimotor control, which are causing great excitement in neuroscience. (These, too, are largely based on Hinton's work.) Higher neural levels send messages downwards predicting the incoming signals from sensors, and only the unpredicted 'error' messages are sent upward. Repeated cycles of this type fine-tune the predicting networks, so that they gradually learn what to expect. Researchers speak of 'the Bayesian brain', because the predictions can be interpreted in terms of—and in computer models are actually based in—Bayesian statistics (see Chapter 2).

Compared with the brain, ANNs are too neat, too simple, too few, and too dry. Too neat, because human-built networks prioritize mathematical elegance and power, whereas biologically evolved brains do not. Too simple, because a single neuron—of which there are about thirty different types—is as computationally complex as an entire PDP system, or even a small computer. Too few, because even ANNs with millions of units are tiny compared with human brains (see Chapter 7). And too dry, because ANN researchers typically ignore not only temporal factors such as neural spiking frequencies and synchronies, but also the biophysics of dendritic spines, neuromodulators, synaptic currents, and the passage of ions.

Each of those shortcomings is lessening. Increased computer power is enabling ANNs to comprise many more individual units. Hugely more detailed models of single neurons are being built, already addressing the computational functions of all the neurological factors just mentioned. The 'dry-ness' is even decreasing in reality, as well as in simulation (some 'neuromorphic' research combines living neurons with silicon chips). And much as the DQN algorithm simulates processes in the visual cortex and hippocampus (see Chapter 2), so future ANNs will doubtless borrow other functions from neuroscience.

Nevertheless, it remains true that ANNs are unlike brains in countless important ways—some of which we don't yet know.

Network scandal

The excitement on PDP's arrival was due largely to the fact that ANNs (a.k.a. connectionism) had been pronounced a dead end twenty years earlier. As remarked in Chapter 1, that judgement had come in a savage 1960s critique by Marvin Minsky and Seymour Papert, both of whom had stellar reputations within the AI community. By the 1980s, ANNs seemed to be not just a dead end but actually dead. Indeed, cybernetics in general had been marginalized (see Chapter 1). Almost all the research funding had gone into symbolic AI instead.

Some early ANNs had seemed enormously promising. Rosenblatt's self-organizing perceptrons—often watched by mesmerized journalists—could learn to recognize patterns even though they started from a random state. He had made hugely ambitious claims, covering all of human psychology, for the potential of his approach. He had pointed out certain limitations, to be sure. But his intriguing 'convergence proof' had *guaranteed* that simple perceptrons can learn to do anything that it's possible to program them to do. That was strong stuff.

But Minsky and Papert, in the late 1960s, offered proofs of their own. They showed mathematically that simple perceptrons cannot do certain things that one would intuitively expect them to be able to do (and which GOFAI could do easily). Their proofs—like Rosenblatt's convergence theorem—applied only to single-layer networks. But their 'intuitive judgment' was that multilayer systems would be defeated by the combinatorial explosion. In other words, perceptrons wouldn't scale up.

Most AI scientists were persuaded that connectionism could never succeed. A few people carried on with ANN research regardless. Indeed, some highly significant progress was made on analysing associative memory (by Christopher Longuet-Higgins and David Willshaw, and later by James Anderson, Teuvo Kohonen, and John Hopfield). But that work was hidden in the background. The groups concerned didn't identify themselves as 'AI' researchers and were generally ignored by those who did.

The arrival of PDP trounced this scepticism. Besides some impressive functioning models (such as the past-tense learner), there were two new convergence theorems: one guaranteeing that a PDP system based on the Boltzmann equations of thermodynamics will reach equilibrium (although perhaps after a *very* long time), and the other proving that a three-layer network can in principle solve any problem presented to it. (*Health warning*: as is also the case in symbolic AI, representing a problem in a way that can be input to the computer is often the most difficult part of the exercise.) Naturally, excitement ensued. The consensus in mainstream AI was shattered.

Symbolic AI had assumed that effortless intuitive thinking is just like conscious inference, but without the consciousness. Now, the PDP researchers were saying that these are fundamentally different kinds of thought. The leaders of the PDP movement (David Rumelhart, Jay McClelland, Donald Norman, and Hinton) all pointed out that *both* types are key to human psychology. But the PDP propaganda—and the general public's reaction to it—implied that symbolic AI, considered as the study of minds, was a waste of time. The worm had well and truly turned.

AI's main funder, the US Department of Defense, made a U-turn too. After an emergency meeting in 1988, they admitted that their previous neglect of ANNs had been 'undeserved'. Now, PDP research was showered with money.

As for Minsky and Papert, they were unrepentant. In the 2nd edition of their anti-ANN book, they allowed that 'the future of network-based learning machines [is] rich beyond imagining'. However, they insisted that high-level intelligence cannot arise from pure randomness, nor from a wholly non-sequential system. Accordingly, the brain must sometimes act as a serial processor, and human-level AI will have to employ hybrid systems. They protested that their critique wasn't the only factor that had led ANNs into their wilderness years: for one thing, computer power had been insufficient. And they denied that they had been trying to divert the research money into symbolic AI. In their words, 'We did not think of our work as killing Snow White; we saw it as a way to understand her'.

Those were respectable scientific arguments. But their initial critique had dripped with vitriol. (The draft was even more venomous: friendly colleagues persuaded them to tone it down, to give the scientific points more prominence.) It's not surprising that it sparked emotion. The persevering ANN devotees deeply resented their new-found cultural invisibility. The furore caused by PDP was even greater. The 'death' and renaissance of ANNs involved jealousy, spite, self-aggrandizement, and gleeful gloating: '*We told you so!*'

This episode was a prime example of a scientific scandal—and not the only one to arise in AI. Theoretical disagreements were embroiled with personal emotions and rivalries, and disinterested thinking was rare. Bitter insults hit the air, and the presses too. AI isn't a passionless affair.

Connections aren't everything

Most accounts of ANNs imply that the only important thing about a neural network is its anatomy. *Which* units are linked to *which others*, and *how strong* are the weights? Certainly, those questions are crucial. However, recent neuroscience has shown that biological circuits can sometimes *alter* their computational function (not merely make it more or less probable), due to chemicals diffusing through the brain.

Nitrous oxide (NO), for example, diffuses in all directions, and its effects—which depend on the concentration at relevant points—endure until it decays. (The rate of decay can be varied by enzymes.) So NO works on all the cells within a given volume of cortex, *whether they're synaptically connected or not*. The functional dynamics of the neural systems concerned are very different from 'pure' ANNs, for volume signalling replaces point-to-point signalling. Analogous effects have been found for carbon monoxide and hydrogen sulphide, and for complex molecules such as serotonin and dopamine.

'So much for ANNs!' an AI sceptic might say. 'There's no chemistry inside computers!' 'It follows', they may add, 'that AI can't model moods, or emotions. For these depend on hormones and neuromodulators'. That very objection was voiced by the psychologist Ulric Neisser in the early 1960s, and some years later by the philosopher John Haugeland in his influential critique of 'cognitivism'. AI might model reasoning, they said, but never affect.

However, these neuroscientific findings have inspired some AI researchers to design ANNs of a radically new type, where linkage *isn't* all. In GasNets, some nodes scattered across the network can release simulated 'gases'. These are diffusible, and modulate the intrinsic properties of other nodes and connections in various ways, depending on concentration. The size of the diffusion volume matters, as does the shape of the source (modelled as a hollow sphere, not a point source). So, a given node will behave differently at different times. In certain gaseous conditions, a node will affect another despite there being no direct link. It's the *interaction* between the gas and the electrical connectivities within the system that is crucial. And, since the gas is emitted only on certain occasions, and diffuses and decays at varying rates, this interaction is dynamically complex.

The GasNet technology was used, for example, to evolve 'brains' for autonomous robots. The researchers found that a specific behaviour might involve two *unconnected* subnets, which worked together because of the modulatory effects. They found, also, that an 'orientation-detector' able to use a cardboard triangle as a navigation aid could evolve in the form of partially *unconnected* sub-networks. They had previously evolved a wholly connected network to do this (see Chapter 5), but the neuromodulatory version evolved more quickly and was more efficient.

So some ANN researchers have moved from considering only anatomy (connections) to recognizing neurochemistry as well.

Different learning rules, and their temporal interactions, can now be simulated with neuromodulation in mind.

Neuromodulation is an analogue phenomenon, not a digital one. Continuously varying concentrations of diffusing molecules are important. Increasingly, AI researchers (using special VLSI chips) are designing networks that combine analogue and digital functions. The analogue features are modelled on the anatomy and physiology of biological neurons, including the passage of ions through the cell membrane. Such 'neuromorphic' computing is being used, for instance, to simulate aspects of perception and motor control. Some AI scientists plan to use neuromorphic computing within 'whole-brain' modelling (see Chapter 7).

Others go even further: instead of modelling ANNs purely *in silico*, they build (or evolve: see Chapter 5) networks comprised of both miniature electrodes and real neurons. For example, when electrodes X and Y are both stimulated artificially, the resulting activity in the 'wet' network results in the firing of some other electrode, Z—thus implementing an *and-gate*. This type of computing (envisaged by Donald Mackay in the 1940s) is in its infancy. But it's potentially exciting.

Hybrid systems

The analogue/digital and hardware/wetware networks just mentioned might understandably be described as 'hybrid' systems. But this term is normally used to refer to AI programs that encompass both symbolic and connectionist information processing.

These had been said by Minsky, in his 1956 manifesto, to be probably necessary, and a few early symbolic programs did combine sequential and parallel processing. But such attempts were rare. As we've seen, Minsky continued to recommend symbolic/ANN hybrids after the arrival of PDP. However, such systems didn't follow immediately (although Hinton built networks combining localist and distributed connectionism, to represent part/whole hierarchies such as family trees).

Indeed, the integration of symbolic and neural-network processing is still uncommon. The two methodologies, logical and probabilistic, are so different that most researchers have expertise only in one.

Nevertheless, some genuinely hybrid systems have been developed, wherein control is passed between symbolic and PDP modules as appropriate. So, the model draws on the complementary strengths of both approaches.

Examples include the Atari game-playing algorithms developed by *DeepMind* (see Chapter 2). These combine deep learning with GOFAI to learn how to play a visually diverse suite of computer games. They use reinforcement learning: no handcrafted rules are provided, only the input pixels and the numerical scores at each step. Many possible rules/plans are considered simultaneously, and the most promising decides the next action. (Future versions will focus on 3D games such as *Minecraft*, and on applications such as driverless cars.)

Other examples are the whole-mind systems ACT-R* and CLARION (see Chapter 2) and LIDA (see Chapter 6). These are deeply informed by cognitive psychology, having been developed for scientific, not technological, purposes.

Some hybrid models take account of specific aspects of neurology, too. For example, the clinical neurologist Timothy Shallice, with the PDP pioneer Norman, published a hybrid theory of familiar ('overlearned') action in 1980, which was later implemented. The theory explains certain common errors. For instance, stroke patients often forget that the letter should be put into the envelope before the sticky flap is licked; or they may get into bed on going upstairs to change their clothes, or pick up the kettle instead of the teapot. Similar errors—of *order, capture*, and *object substitution*—occur occasionally in all of us.

But why? And why are brain-damaged patients especially prone to them? Shallice's computational theory claims that familiar action is generated by two types of control, which can break down or take over at specific points. One, 'contention scheduling', is automatic. It involves (unconscious) competition between various hierarchically organized action schemata. Control goes to the one whose activation has exceeded some threshold. The other ('executive') control mechanism is conscious. It involves the deliberative supervision and modulation of the first mechanism—including planning, and repairing mistakes. For Shallice, contention scheduling is modelled by PDP, executive control by symbolic AI.

The activation level of an action schema can be raised by perceptual input. For instance, one's unthinking glimpse (pattern recognition) of the bed, on reaching the bedroom, can trigger the action schema of getting into bed, even though the original intention (plan) had been to change one's clothes.

Shallice's theory of action was initiated by using ideas from AI (notably, models of planning), which resonated with his own clinical experience. It was later supported by evidence from brain scanning. And recent neuroscience has discovered other factors, including neurotransmitters, implicated in human action. These are now represented in the current computer models based on the theory.

Interactions between contention scheduling and executive control are relevant also to robotics. An agent following a plan should be able to halt or vary it, depending on what it observes in the environment. That strategy characterizes robots that combine *situated* and *deliberative* processing (see Chapter 5).

Anyone interested in AGI should note that those few AI scientists who have seriously considered the computational architecture of *the mind as a whole* accept hybridism unreservedly. They include Allen Newell and Anderson (whose SOAR and ACT* were discussed in Chapter 2), Stan Franklin (whose LIDA model of consciousness is outlined in Chapter 6), Minsky (with his 'society' theory of mind), and Aaron Sloman (whose simulation of anxiety is described in Chapter 3).

In short, the virtual machines implemented in our brains are both sequential and parallel. Human intelligence requires subtle cooperation between them. And human-level AGI, if it's ever achieved, will do so too.

Chapter 5
Robots and artificial life

A-Life models biological systems. Like AI in general, it has both technological and scientific aims. A-life is integral to AI, because all the intelligence we know about is found in living organisms. Indeed, many people believe that mind can arise *only* from life (see Chapter 6). Hard-headed technologists don't worry about that question. But they do turn to biology in developing practical applications of many kinds. These include robots, evolutionary programming, and self-organizing devices. Robots are quintessential examples of AI: they have high visibility and are hugely ingenious—and very big business, too. Evolutionary AI, although widely used, is less well known. Self-organizing machines are even less familiar (unsupervised learning excepted: see Chapter 4). Nevertheless, in the quest to understand self-organization, AI has been as useful to biology as biology has been to AI.

Situated robots and interesting insects

Robots were built centuries ago—by Leonardo da Vinci, among others. AI versions emerged in the 1950s. William Grey Walter's post-war 'tortoises' amazed observers by avoiding obstacles and finding light. And a main aim of MIT's newly founded AI Laboratory was to build 'the MIT robot', integrating computer vision, planning, language, and motor control.

There has been huge advance since then. Now, some robots can climb hills, stairs, or walls; some can run fast, or jump high; and some can carry—and throw—heavy burdens. Others can break themselves up and re-assemble the parts, sometimes adopting a new shape—like a worm (able to traverse a narrow pipe), or a ball or multi-legged creature (suited to level or rough ground respectively). What drove that advance was a turn from psychology to biology.

Classical AI robots emulated human voluntary action. Drawing on theories of cerebral modelling, they employed internal representations of the world and of the agent's own actions. But they weren't impressive. Because they relied on abstract planning, they were subject to the frame problem (see Chapter 2). They couldn't react promptly, because even slight environmental changes required anticipatory planning to restart; nor could they adapt to novel (unmodelled) circumstances. Steady movement was difficult even on level, uncluttered ground (hence the SRI robot's nickname: SHAKEY), and fallen robots couldn't recover. They were useless in most buildings—never mind on Mars.

Today's robots are very different. The focus has changed from humans to insects. Insects probably aren't intelligent enough to model the world, or to plan. Yet they manage. Their behaviour—*behaviour*, not *action*—is appropriate, adaptive. But it's mainly reflex rather than deliberate. They respond unthinkingly to the current situation, not to some imagined possibility or goal state. Hence the labels: 'situated', or 'behaviour-based', robotics. (Situated behaviour isn't confined to insects: social psychologists have identified many situation-bound behaviours in humans.)

In trying to give comparable reflexes to AI machines, roboticists favour engineering over programming. If possible, sensorimotor reflexes are physically embodied in the robot's anatomy, not provided as software code.

How far robot anatomy should match the anatomy of living organisms is debatable. For technological purposes, ingenious engineering tricks are acceptable. Today's robots incorporate many unrealistic gimmicks. But perhaps biological mechanisms are especially efficient? They are certainly adequate. So roboticists also consider real animals: what they can do (including their various navigational strategies), what sensory signals and specific movements are involved, and what neurological mechanisms are responsible. The biologists, in turn, employ AI modelling to investigate these mechanisms: a research field named *computational neuroethology*.

One example is the cockroach robotics of Randall Beer. Cockroaches have six multi-segmented legs—suggesting both advantage and disadvantage. Hexopod locomotion is more stable than bipedalism (and more generally useful than wheels). However, coordinating six limbs appears more difficult than coordinating two. Besides deciding which leg should be moved next, the creature must find the correct placement, force, and timing. And how should the legs interact? They must be largely independent, because there could be a pebble near only one leg. But if that leg is lifted higher, the others must compensate to retain the balance.

Beer's robots reflect the neuroanatomy and sensorimotor controls of real cockroaches. They can climb stairs, walk on rough ground, clamber over obstacles (instead of merely avoiding them), and recover from falling over.

The roboticist Barbara Webb looks at crickets, not cockroaches. Her focus isn't on locomotion (so her robots can use wheels). Rather, she wants her devices to identify, locate, and approach a particular sound pattern. Clearly, such behaviour ('phonotaxis') could have many practical applications.

Female crickets can do this on hearing the song of a conspecific male. However, the cricket can recognize only one song, sung at only one speed and frequency. Speed and frequency vary for different species of cricket. But the female doesn't *choose between* different songs, for she doesn't possess feature detectors coding a range of sounds. She uses a mechanism that's sensitive only to one frequency. This isn't a *neural* mechanism, like the auditory detectors in human brains. It's a fixed-length tube in her thorax, connected to the ears on her front legs and to her spiracles. The length of the tube is an exact proportion of the wavelength of the male's song. The *physics* ensures that phase cancellations (between the air in the tube and the air outside) occur only for songs with the right frequency, and that the intensity difference depends wholly on the direction of the sound source. The female insect is neurally hardwired to move in that direction: he sings, she goes. Situated behaviour, indeed.

Webb chose cricket phonotaxis because it had been closely studied by neuroethologists. But there were many unanswered questions: whether (and how) the song's direction and sound are processed independently; whether identifying and locating it are independent; how the female's walking is triggered; and how its zigzagging direction is controlled. Webb devised the simplest possible mechanism (only four neurons) that could generate similar behaviour. Later, her model used more neurons (based on detailed real-life data); included additional neural features (e.g. latency, firing rate, and membrane potential); and integrated hearing with vision. Her work has clarified many neuroscientific questions, answered some, and raised more. So it's been helpful for biology, as well as robotics.

(Although robots are physical things, much robotics research is done in simulation. Beer's robots, for instance, are sometimes *evolved in software* before being built. Similarly, Webb's are *designed as programs* before being tested in the real world.)

Despite the turn to insects in mainstream robotics, research on android robots continues. Some are mere toys. Others are 'social', or 'companion', robots, designed for home use by elderly and/or disabled people (see Chapter 3). These are intended less as fetch-and-carry slaves than as autonomous personal assistants. Some appear 'cute', having long eyelashes and seductive voices. They can make eye contact with users, and recognize individual faces and voices. Also, they can—up to a point—hold unscripted conversations, interpret the user's emotional state, and generate 'emotional' responses (human-like facial expressions and/or speech patterns) themselves.

Although some robots are large (for handling heavy loads and/or traversing rough ground), most are small. Some—for use inside blood vessels, for example—are very small. Often, they are sent to work in large numbers. Whenever multiple robots are involved in a task, questions arise about how (if at all) they communicate, and how that enables the group to do things that couldn't be done individually.

For answers, roboticists often consider social insects, like ants and bees. Such species exemplify 'distributed cognition' (see Chapter 2), in which knowledge (and appropriate action) is spread across an entire group rather than being available to any one animal.

If the robots are extremely simple, their developers may speak of 'swarm intelligence', and analyse cooperative robot systems as cellular automata (CAs). A CA is a system of individual units, each taking one of a finite number of states by following simple rules, which depend on the current state of its neighbours. The overall pattern of a CA's behaviour may be surprisingly complex. The basic analogy is living cells cooperating in multicellular organisms. The many AI versions include the flocking algorithms used for crowds of bats or dinosaurs in Hollywood animations.

The concepts of distributed cognition and swarm intelligence apply also to human beings. The latter is used when the 'knowledge' concerned isn't something which any participating individual can possess (e.g. the overall behaviour of large crowds). The former is more often used when the participating individuals *could* possess all the relevant knowledge, but don't. For instance, the anthropologist Edwin Hutchins has shown how knowledge of navigation is shared across a ship's crew members—and also embodied in physical objects, such as charts and (the location of) chart tables.

To speak of knowledge as being embodied in physical objects may seem strange, or at best metaphorical. But there are many today who claim that human minds are *literally* embodied, not only in people's physical actions but also in the cultural artefacts they engage with in the external world. This 'external/embodied mind' theory is partly grounded in work done by the leader of the human-to-insect switch in robotics: MIT's Rodney Brooks.

Brooks is now a prime developer of robots for the US military. In the 1980s, he was a fledgling roboticist frustrated by the impracticality of symbolic AI's world-modelling planners. He turned to situated robotics for purely technological reasons, but soon developed his approach into a theory about adaptive behaviour in general. This went far beyond insects: even human action, he argued, doesn't involve internal representations. (Or, he sometimes implied, doesn't *usually* involve representations.)

His critique of symbolic AI excited the psychologists and philosophers. Some were highly sympathetic. Psychologists had already pointed out that much human behaviour is situation-bound: role-playing in distinct social environments, for example. And cognitive psychologists had highlighted animate vision, in which the agent's own bodily movement is key. Today, theories of embodied mind are hugely influential outside AI (see Chapter 6).

But some others, such as David Kirsh, were—and still are—vehemently opposed, arguing that compositional representations are necessary for those types of behaviour which involve concepts. For example, recognizing perceptual invariance, in which an object can be recognized from many different views; re-identifying individuals over time; anticipatory self-control (planning); negotiating, not merely scheduling, conflicting motives; counterfactual reasoning; and language. These critics admit that situated robotics shows concept-free behaviour to be more widespread than many philosophers believe. Nevertheless, logic, language, and thoughtful human action all require symbolic computation.

Many roboticists, too, reject Brooks's more extreme claims. Alan Mackworth's group, one of several working on robot soccer, refers to 'reactive deliberation'—which includes sensory perception, real-time decision making, planning, plan recognition, learning, and coordination. They are seeking an *integration* of GOFAI and situated perspectives. (That is, they're building *hybrid* systems: see Chapter 4.)

In general, representations are critical for the process of action selection in robotics, but less so for the execution of actions. So, the jokers who said that 'AI' now stands for 'artificial insects' weren't quite right.

Evolutionary AI

Most people assume that AI necessitates meticulous design. Given the unforgiving nature of computers, how could it be otherwise? Well, it can.

Evolutionary robots (which include some situated robots), for instance, result from a combination of rigorous programming/ engineering and random variation. They are unpredictably evolved, not carefully designed.

Evolutionary AI in general has this character. It was initiated within symbolic AI, but is also used in connectionism. Its many practical applications include art (where unpredictability may be welcome) and the development of safety-critical systems such as aircraft engines.

A program can change itself (instead of being rewritten by a programmer), and can even improve itself, by using *genetic algorithms* (GAs). Inspired by real-life genetics, these enable both random variation and non-random selection. The selection requires a criterion of success, or 'fitness function' (analogous to natural selection in biology) working alongside the GAs. Defining the fitness function is crucial.

In evolutionary software, the initial task-oriented program can't solve the task efficiently. It may not be able to solve it at all, for it may be an incoherent collection of instructions or a randomly connected neural network. But the overall program includes GAs in the background. These can change the task-oriented rules. The changes, made randomly, resemble point mutation and crossover in biology. So a single symbol in a programmed instruction may be altered, or short symbol sequences may be 'swapped' across two instructions.

The various task programs within any one generation are compared, and the most successful are used to breed the next generation. A few (randomly chosen) others may be retained also, so that potentially useful mutations that haven't yet had any good effect aren't lost. As the generations pass, the task program's efficiency increases. Sometimes, the *optimal* solution is found. (In some evolutionary systems the credit-assignment problem—see Chapter 4—is solved by some variant of John Holland's 'bucket-brigade' algorithm, which identifies *just which* parts of a complex evolutionary program are most responsible for its success.)

Some evolutionary AI is fully automatic: the program applies the fitness function at every generation, and is left to evolve unsupervised. Here, the task must be very clearly defined—by the physics of aircraft engines, for example. Evolutionary art, by contrast, is usually highly interactive (the *artist* selects the best at every generation), because the fitness function—aesthetic criteria—can't be stated clearly.

Most evolutionary robotics is intermittently interactive. The robot's anatomy (e.g. sensors and sensorimotor connections) and/or its controller ('brain') evolve automatically, but *in simulation*. For most generations, no physical robot exists. But at every 500th generation, perhaps, the evolved design is tested in a physical device.

Useless mutations tend not to survive. Inman Harvey's team at the University of Sussex found (in 1993) that one of a robot's two 'eyes', and all of its 'whiskers', may lose their initial connections to the controlling neural network if the task needs neither depth vision nor touch. (Similarly, auditory cortex in the congenitally deaf, or in animals deprived of auditory input, gets used for visual computation: the brain evolves *within a lifetime*, not only across generations.)

Evolutionary AI can provide deep surprises. For instance, a situated robot being evolved (also at Sussex) to generate obstacle-avoiding movement towards a goal developed an orientation detector analogous to those found in brains. The robot's world included a white cardboard triangle. Unexpectedly, a randomly connected mini-network arose in the controller which responded to a light/dark gradient at a particular orientation (one side of the triangle). This then evolved as an integral part of a visuo-motor mechanism, its (initially random) connections to motor units enabling the robot to use the object as a navigation aid. The mechanism didn't work for a black triangle, nor for the opposite side. And it was a stand-alone item, there being no comprehensive *system* of orientation detectors. It was useful, nonetheless. This startling result was broadly repeatable. Using neural nets of different types, the Sussex team found that every successful solution evolved some active orientation detector—so the high-level behavioural strategy was the same. (The *exact* implementation details varied, but were often very similar.)

On another occasion, GAs were being used by the Sussex team to design hardware electrical circuits. The task was to evolve oscillators. However, what emerged was a primitive radio-wave sensor, picking up the background signal from a nearby PC monitor. This depended on unforeseen physical parameters. Some were predictable (the aerial-like properties of all printed circuit boards), although not previously considered by the team. But others were accidental, and seemingly irrelevant. These included spatial proximity to a PC monitor; the order in which the analogue switches had been set; and the fact that a soldering iron left on a nearby workbench was plugged into the mains. (This result wasn't repeatable: next time, the radio antenna might be influenced by the chemistry of the wallpaper.)

The radio-wave sensor is interesting because many biologists (and philosophers) argue that nothing radically new could emerge from AI, since all the results of a computer program (including the random effects of GAs) must lie within the space of possibilities defined by it. Only biological evolution, they say, can generate new perceptual sensors. They allow that a feeble AI visual sensor could evolve into a better one. But the *very first* visual sensor, they say, could emerge only in a physical world governed by causation. A random genetic mutation causing a light-sensitive chemical could introduce light, already present in the *outside world*, into the organism's *environment*. However, the unexpected radio sensor *similarly* brought radio waves into the device's 'environment'. It did depend partly on physical causation (plugs, etc.). But it was an exercise in AI, not biology.

Radical novelty in AI does indeed require outside influences, because it's true that a program cannot surpass its possibility space.

But these influences needn't be physical. A GA system connected to the Internet might evolve fundamental novelties by interacting with a virtual world.

Another, much earlier, surprise within evolutionary AI instigated still ongoing research into evolution as such. The biologist Thomas Ray used GAs to simulate the ecology of tropical rainforests. He saw the spontaneous emergence of parasites, resistance to parasites, and super-parasites capable of overcoming that resistance. He also discovered that sudden 'leaps' in (phenotypic) evolution can be generated by a succession of tiny underlying (genotypic) mutations. Orthodox Darwinians already believed this, of course. But it's so counter-intuitive that some biologists, such as Stephen Jay Gould, had argued that non-Darwinian processes must also be involved.

Today, simulated mutation rates are being systematically varied and traced, and GA researchers are analysing 'fitness landscapes', 'neutral [*sic*] networks', and 'genetic drift'. This work explains how mutations can be preserved even though they haven't (yet) increased reproductive fitness. So AI is helping biologists to develop evolutionary theory in general.

Self-organization

The key feature of biological organisms is their ability to structure themselves. Self-organization is the spontaneous emergence of order from an origin that's ordered to a lesser degree. It's a puzzling, even quasi-paradoxical, property. And it's not obvious that it could happen in non-living things.

Broadly speaking, self-organization is a creative phenomenon. Psychological creativity (both 'historical' and 'individual') was discussed in Chapter 3, and self-organized (unsupervised) associative learning in Chapter 4. Here, our focus is on the types of self-organization studied in biology.

Examples include phylogenetic evolution (a form of historical creativity); embryogenesis and metamorphosis (analogous to individual creativity in psychology); brain development (individual creativity followed by historical creativity); and cell formation (historical creativity when life began, individual creativity thereafter). How can AI help us to understand these?

Alan Turing explained self-organization in 1952 by going back to basics. He asked how something homogeneous (such as the undifferentiated ovum) could originate structure. He acknowledged that most biological development adds new order to pre-existing order: the sequence of changes in the embryo's neural tube, for instance. But order-from-homogeneity is the fundamental (and mathematically simplest) case.

Embryologists had already posited 'organizers': unknown chemicals directing development in unknown ways. Turing couldn't identify the organizers either. Instead, he considered highly general principles about chemical diffusion.

He showed that, if different molecules met, the results would depend on their rates of diffusion, their concentrations, and the speeds at which their interactions would destroy/construct molecules. He did this by varying the numbers in imaginary chemical equations and investigating the results. Some number combinations produced only formless mixtures of chemicals. But others generated order—for example, regular peaks of concentration of a certain molecule. Such chemical peaks, he said, might be biologically expressed as surface markings (stripes), or as the origins of repeated structures such as petals or bodily segments. Diffusion reactions in three dimensions could produce hollowing-out, like gastrulation in the early embryo.

These ideas were immediately recognized as hugely exciting. They solved the previously intractable puzzle of how order can arise from an unordered origin. But 1950s biologists couldn't do much with them. Turing had relied on mathematical analysis. He did do some (grindingly tedious) simulation by hand, followed by modelling on a primitive computer. But his machine didn't have enough computational power to do the relevant sums, or to explore number variations systematically. Nor were computer graphics available, to convert lists of numbers into visibly intelligible form.

Both AI and biology had to wait forty years before Turing's insights could be developed. The computer-graphics expert Greg Turk explored Turing's own equations, sometimes 'freezing' the results of one equation before applying another. This procedure, reminiscent of the on/off switching of genes, exemplified pattern-from-pattern—which Turing had mentioned, but couldn't analyse. In Turk's AI model, Turing's equations generated not only Dalmation markings and stripes (as his hand simulations had done), but also leopard spots, cheetah spots, giraffe reticulations, and lion fish patterns.

Other researchers used more complicated sequences of equations, getting more complex patterns accordingly. Some were developmental biologists, who now know more about the actual biochemistry.

For instance, Brian Goodwin studied the life cycle of the alga *Acetabularia*. This unicellular organism transforms itself from a shapeless blob into an elongated stalk; it then grows a flattened top; next, it develops a ring of knobs around the edge; these later sprout into a whorl of laterals, or branches; finally, the laterals coalesce to form an umbrella-shaped cap. Biochemical experiments show that over thirty metabolic parameters are involved (e.g. calcium concentrations; the affinity between calcium and certain proteins; and the mechanical resistance of the cytoskeleton). Goodwin's computer model of *Acetabularia* simulated complex, iterative feedback loops in which these parameters can change from moment to moment. Various bodily metamorphoses resulted.

Like Turing and Turk, Goodwin played around with numerical values to see which ones would actually generate new forms. He used only numbers within the ranges observed in the organism, but these were randomized.

He found that certain patterns—for example, alternating high/low concentrations of calcium at the tip of a stalk (the emerging symmetry of a whorl)—arose repeatedly. They didn't depend on a particular choice of parameter values, but emerged spontaneously if the values were set anywhere within a large range. Moreover, once the whorls had originated, they persisted. So, said Goodwin, they might become the ground for transformations leading on to other frequently occurring features. This could happen in phylogenesis as well as ontogenesis (historical creativity as well as individual creativity)—in the evolution of the tetrapod limb, for instance.

No umbrella cap was ever generated by this model. Possibly, that would require extra parameters, representing as yet unknown chemical interactions within real *Acetabularia*. Or perhaps such caps do lie within the model's possibility space, so could in principle arise from it, but *only* if the numerical values are so strictly limited that they're unlikely to be found by random search. (The laterals weren't generated either, but that was due simply to lack of computational power: the whole program would need to be executed on a lower level, for every individual lateral.)

Goodwin drew an intriguing theoretical moral. He saw whorls as 'generic' forms, occurring—unlike umbrella caps—in many animals and plants. This suggests that they're due not to highly specific biochemical mechanisms directed by contingently evolved genes, but rather to general processes (like reaction diffusion) found in most or even all living things. Such processes might form the basis of a 'structuralist' biology: a general science of morphology, whose explanations would be prior to, although fully consistent with, Darwinian selection. (This possibility was implied by Turing's discussion, and had been stressed by D'Arcy Thompson, a biologist he had cited; but Turing himself ignored it.)

Reaction diffusion works by physico-chemical laws determining local molecular interactions—that is, by laws representable in cellular automata. When John von Neumann defined CAs, he pointed out that they are in principle applicable to physics. Today's A-Life researchers use CAs for many purposes, the generation of biological patterns being particularly relevant here. For example, very simple CAs, defined on only one dimension (a line), can generate remarkably lifelike patterns—like those on seashells, for instance.

Especially intriguing, perhaps, is A-Life's use of CAs in attempting to describe 'life as it could be', not only 'life as we know it'. Christopher Langton (who named 'artificial life' in 1987) explored numerous randomly defined CAs, noting their propensity to generate order. Many produced only chaos. Others formed boringly repetitive, or even static, structures. But a few generated subtly changing yet relatively stable patterns—characteristic, Langton said, of living things (and of computation, too). Surprisingly, these CAs shared the same numerical value on a simple measure of the system's informational complexity. Langton suggested that this 'lambda-parameter' applies to all possible living things, whether on Earth or Mars.

Self-organization moulds not only whole bodies, but also organs. The brain, for example, develops by evolutionary processes (within a lifetime and across generations), as well as by unsupervised learning. Such learning can have highly idiosyncratic (historically creative) results. But early cerebral development in each individual also creates *predictable* neural structures.

For instance, newborn monkeys possess orientation detectors systematically spanning 360 degrees. These can't have been learned from experience of the external world, so it's natural to assume that they are coded in the genes. But they aren't. Instead, they arise spontaneously from an initially random network.

This has been shown not only by biologically realistic computer modelling done by neuroscientists, but also by 'pure' AI. The IBM researcher Ralph Linsker has defined multilayer feedforward networks (see Chapter 4) showing that simple Hebbian rules, given *random* activity (such as 'noise' within the embryonic brain), can generate structured collections of orientation detectors.

Linsker doesn't rely on practical demonstrations alone, nor focus only on orientation detectors: his abstract 'infomax' theory is applicable to *any* network of this type. It states that network connections develop to maximize the amount of information preserved when signals are transformed at each processing stage. All connections form under certain empirical constraints, such as biochemical and anatomical limitations. However, the mathematics guarantees that a cooperative system of communicating units will emerge. The infomax theory relates to phylogenetic evolution, too. It makes it less counter-intuitive that a single mutation, in the evolution of a complex system, will be adaptive. The apparent need for several *simultaneous* mutations evaporates if each level can spontaneously adapt to a small alteration in another.

As regards self-organization at the cellular level, both intracellular biochemistry and the formation of cells/cell walls have been modelled. This work exploits that of Turing on reaction diffusion. However, it relies more on biological concepts than on ideas originated within A-Life.

In sum, AI provides many theoretical ideas regarding self-organization. And self-organizing artefacts abound.

Chapter 6
But is it intelligence, really?

Suppose that future AGI systems (on-screen or robots) equalled human performance. Would they have *real* intelligence, *real* understanding, *real* creativity? Would they have selves, moral standing, free choice? Would they be conscious? And without consciousness, could they have any of those other properties?

These aren't scientific questions, but philosophical ones. Many people feel intuitively that the answer, in each case, is '*Obviously*, no!'

Things aren't so straightforward, however. We need careful arguments, not just unexamined intuitions. But such arguments show that there are no unchallengeable answers to these questions. That's because the concepts involved are themselves highly controversial. Only if they were all satisfactorily understood could we be *confident* that the hypothetical AGI would, or wouldn't, really be intelligent. In short: no one knows for sure.

Some might say it doesn't matter: what the AGIs will actually do is what's important. However, our answers could affect *how we relate to them*, as we'll see.

This chapter, then, won't provide unequivocal answers. But it will suggest that some answers are more reasonable than others. And it will show how AI concepts have been used by (some) philosophers to illuminate the nature of *real* minds.

The Turing Test

In a paper published in the philosophy journal *Mind* in 1950, Alan Turing described what's called the Turing Test. This asks whether someone could distinguish, 30 per cent of the time, whether they were interacting (for up to five minutes) with a computer or a person. If not, he implied, there'd be no reason to deny that a computer could really think.

That was tongue in cheek. Although it featured in the opening pages, the Turing Test was an adjunct within a paper primarily intended as a manifesto for a future AI. Indeed, Turing described it to his friend Robin Gandy as light-hearted 'propaganda', inviting giggles rather than serious critique.

Nevertheless, the philosophers pounced. Most argued that even if a program's responses were indistinguishable from a human's, this wouldn't *prove* its intelligence. The most common objection was—and still is—that the Turing Test concerns only observable behaviour, so could be passed by a zombie: something behaving exactly like us, but lacking consciousness.

This objection assumes that intelligence requires consciousness, and that zombies are logically possible. We'll see (in the section 'AI and phenomenal consciousness') that some accounts of consciousness imply that the concept of *zombie* is incoherent. If they are right, then no AGI could be a zombie. In that respect, the Turing Test would be justified.

The Turing Test interests philosophers (and the general public) greatly. But it hasn't been important in AI. Most AI aims to provide useful tools, not to mimic human intelligence—still less, to make users believe they're interacting with a person.

Admittedly, publicity-hungry AI researchers sometimes claim, and/or allow journalists to claim, that their system passes the Turing Test. However, these tests don't fit Turing's description. For instance, Ken Colby's model PARRY 'fooled' psychiatrists into thinking that they were reading interviews with paranoiacs—because *they naturally assumed* that they were dealing with human patients. Similarly, computer art is often ascribed to human beings *if* there's no hint that a machine might be involved.

The closest thing to a genuine Turing Test is the annual Loebner competition (now held in Bletchley Park). The current rules prescribe twenty-five-minute interactions, using twenty pre-selected questions designed to test memory, reasoning, general knowledge, and personality. The judges consider relevance, correctness, and the clarity and plausibility of expression/grammar. As yet, no program has fooled the Loebner judges for 30 per cent of the time. (In 2014, a program said to be a 13-year-old Ukrainian boy deceived 33 per cent of its interrogators; but mistakes are readily forgiven in non-native speakers, especially children.)

The many problems of consciousness

There's no such thing as *the* problem of consciousness. Rather, there are many. The word 'conscious' is used to make many different distinctions: awake/asleep; deliberate/unthinking; in/out of attention; accessible/inaccessible; reportable/non-reportable; self-reflective/unexamined; and so on. No single explanation will clarify all of them.

The contrasts just listed are *functional* ones. Many philosophers would allow that they could in principle be understood in information-processing and/or neuroscientific terms.

But *phenomenal* consciousness—sensations (like blueness, or pain), or 'qualia' (the philosophers' technical term)—seems to be different. The very existence of qualia, in a basically material universe, is a notorious metaphysical puzzle.

David Chalmers calls this 'the hard problem'. And, he says, it's inescapable: '[We must] take consciousness seriously.... [To] redefine the problem as that of explaining *how certain cognitive or behavioural functions are performed* is unacceptable'.

Various highly speculative solutions have been suggested. These include Chalmers's own version of pan-psychism, a self-confessedly 'outrageous, or even crazy' theory according to which phenomenal consciousness is an irreducible property of the universe, analogous to mass or charge. Several other theorists have appealed to quantum physics—using one mystery to solve another, their opponents say. Colin McGinn has even argued that humans are constitutionally incapable of understanding the causal link between brain and qualia, much as dogs cannot understand arithmetic. And Jerry Fodor, a leading philosopher of cognitive science, believes that: 'Nobody has the slightest idea how anything material could be conscious. Nobody even knows what it would be like to have the slightest idea how anything material could be conscious'.

In short, very few philosophers claim to understand phenomenal consciousness—and those who do are believed by almost nobody else. The topic is a philosophical morass.

Machine consciousness

Thinkers who are sympathetic to AI approach consciousness in two ways. One is building computer models of consciousness: this is called 'machine consciousness' (MC). The other (which is characteristic of AI-influenced philosophers) is *analysing* it in broadly computational terms, without doing modelling.

A *truly* intelligent AGI would possess functional consciousness. For example, it would focus on (pay attention to, be aware of) different things at different times. A human-level system would be able also to deliberate, and self-reflect. It could generate creative ideas, and even deliberately evaluate them. Without those capacities, it couldn't generate seemingly intelligent performance.

Phenomenal consciousness may be involved when humans evaluate creative ideas (see Chapter 3). Indeed, many would say that it attends every 'functional' difference. Nevertheless, MC researchers—all of whom consider functional consciousness—usually ignore phenomenal consciousness.

One interesting MC project is LIDA (Learning Intelligent Distribution Agent), developed in Memphis by Stan Franklin's group. This label names two things. One is a *conceptual* model—a verbally expressed computational theory—of (functional) consciousness. The other is a partial, and simplified, *implementation* of that theoretical model.

Both are used for scientific purposes (Franklin's primary aim). But the second also has practical applications. The LIDA implementation can be customized to suit specific problem domains, for example, medicine.

Unlike SOAR, ACT-R, and CYC (see Chapter 2), it's very recent. The first version (built for the US Navy, to organize new jobs for end-of-duty sailors) was released in 2011. The current version covers attention, and its effects on learning in various types of memory (episodic, semantic, and procedural); and sensorimotor control is now being implemented for robotics. But many features, including language, are still missing. (The description that follows concerns the *conceptual* model, irrespective of which aspects are already implemented.)

LIDA is a hybrid system, involving spreading activation and sparse representations (see Chapter 4) as well as symbolic programming. It is based on Bernard Baars's neuropsychological Global Workspace Theory (GWT) of consciousness.

GWT sees the brain as a distributed system (see Chapter 2), in which a host of specialized sub-systems, functioning in parallel, compete for access to working memory (see Figure 2). Items appear there sequentially (the stream of consciousness), but are 'broadcast' to all cortical areas.

If a broadcast item, derived from a sense organ or other sub-system, triggers a response from a certain area, that response may be

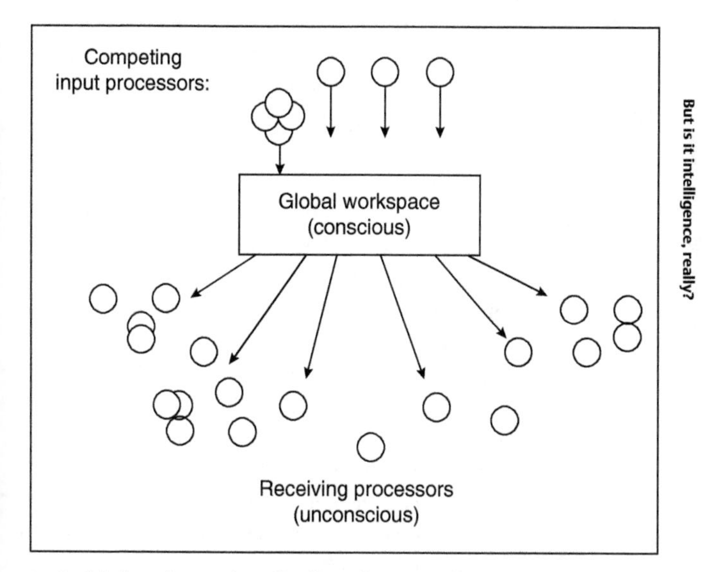

2. A global workspace in a distributed system. The nervous system involves various specialized unconscious processors (perceptual analysers, output systems, planning systems, etc.). Interaction, coordination, and control of these unconscious specialists requires a central information exchange or 'global workspace'. Input specialists can cooperate and compete for access to it. In the case shown here, four input processors cooperate to place a global message, which is then broadcast to the system as a whole.

strong enough to win the competition for attention, which actively controls access to consciousness. (Novel perceptions/representations tend to gain attention, whereas repeated items fade from consciousness.) The sub-systems are often complex. Some are nested hierarchically, and many have associative linkages of diverse kinds. A variety of unconscious contexts (organized in different memories) shape conscious experience, both evoking and amending the items in the global workspace. The contents of attention, in turn, adapt the enduring contexts by causing learning of various types.

These contents, when broadcast, guide the selection of the next action. Many actions are cognitive: building or amending internal representations. Moral norms are stored (in semantic memory) as procedures for appraising potential actions. Decisions can be influenced also by the perceived/predicted reactions of other social agents.

Consider planning, for instance (see Chapter 2). Intentions are represented as largely unconscious but relatively high-level structures, which can lead to conscious goal images (selected by currently salient features from perception, memory, or imagination). These recruit relevant sub-goals. They 'recruit' rather than 'retrieve', for the sub-goals themselves decide their relevance. Like all cortical sub-systems, they lie in wait to be triggered by some broadcast item—here, by an appropriate goal image. LIDA can transform a selected goal-directed action schema into low-level executable motor actions, responsive to detailed features of an unpredictable, and changing, environment.

Baars's theory (and Franklin's version of it) wasn't dreamed up in a computer scientist's workshop. On the contrary, it was designed to take into account a wide variety of well-known psychological phenomena, and a wide range of experimental evidence (see Figure 3). But these authors claim that it also solves some previously unsolved psychological puzzles.

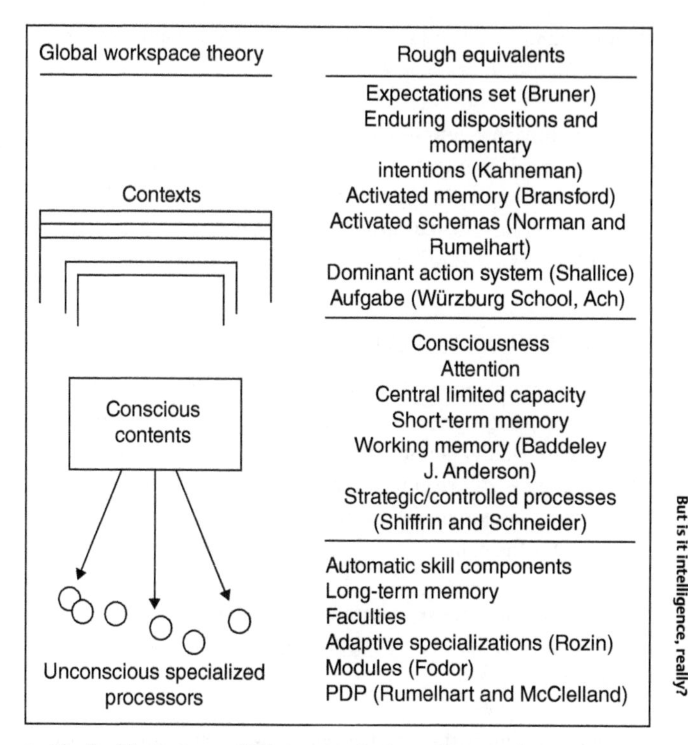

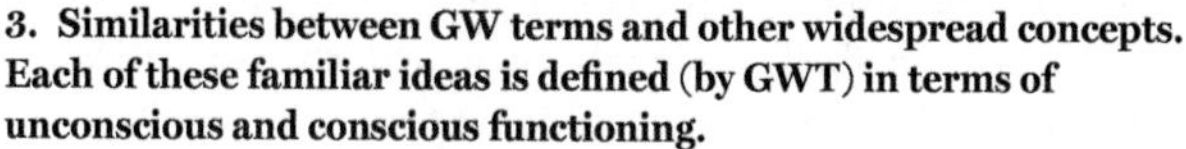
3. Similarities between GW terms and other widespread concepts. Each of these familiar ideas is defined (by GWT) in terms of unconscious and conscious functioning.

For example, they say that GWT/LIDA solves the long-disputed 'binding' problem. This asks how *several* inputs from different senses, in different brain areas—for instance, the feel, appearance, smell, and sound of a cat—are attributed to *one and the same* thing. Franklin and Baars's claim that it also explains how human minds avoid the frame problem (see Chapter 2). When generating creative analogies, for instance, there's no central executive, searching the entire data structure for relevant items. Rather, if a sub-system recognizes that some broadcast item fits/approximates what it's (always) looking for, it competes for entry to the global workspace.

This AI approach is reminiscent of Pandemonium's 'demons', and the 'blackboard' architectures used to implement production systems (see Chapters 1 and 2). That's not surprising, for those ideas inspired Baars's neuropsychological theory, which eventually led to LIDA. The theoretical wheel has turned full circle.

AI and phenomenal consciousness

MC practitioners ignore the 'hard' problem. But three AI-inspired philosophers have addressed it head-on: Paul Churchland, Daniel Dennett, and Aaron Sloman. To say that their answers are controversial is an understatement. Where phenomenal consciousness is concerned, however, that's par for the course.

Churchland's 'eliminative materialism' *denies* the existence of immaterial thoughts and experiences. Instead, he identifies them with brain states.

He offers a scientific theory—part computational (connectionist), part neurological—defining a 4D 'taste-space', which systematically maps subjective discriminations (qualia) of taste onto specific neural structures. The four dimensions reflect the four types of taste receptor on the tongue.

For Churchland, this isn't a matter of mind–brain *correlations*: to have an experience of taste *simply is* to have one's brain visit a particular point in that abstractly defined sensory space. The implication is that *all* phenomenal consciousness is simply the brain's being at a particular location in some empirically discoverable hyperspace. If so, then no computer (possibly excepting a whole-brain emulation) could have phenomenal consciousness.

Dennett, too, denies the existence of ontologically distinct experiences, over and above bodily events. (So a common response to his provocative book of 1991 is: 'Not *Consciousness Explained*, but *explained away*'.)

To experience, in his view, is to discriminate. But in discriminating something that exists in the material world, one doesn't bring something else into existence in some other, immaterial, world. He expresses this in an imaginary conversation:

> [OTTO:] It seems to me that you've denied the existence of the most indubitably real phenomena there are: the real seemings that even Descartes in his *Meditations* couldn't doubt.
>
> [DENNETT:] In a sense, you're right: that's what I'm denying exist. Let's [consider] the neon colour-spreading phenomenon. There seems to be a pinkish glowing ring on the dust jacket. [He's describing a visual illusion, caused by red and black lines on shiny white paper.]
>
> [OTTO:] There sure does.
>
> [DENNETT:] But there isn't any pinkish ring. Not really.
>
> [OTTO:] Right. But there sure seems to be!
>
> [DENNETT:] Right.
>
> [OTTO:] So where is it, then?
>
> [DENNETT:] Where's what?
>
> [OTTO:] The pinkish glowing ring.
>
> [DENNETT:] There isn't any; I thought you'd just acknowledged that.
>
> [OTTO:] Well yes, there isn't any pinkish ring out there on the page, but there sure seems to be.
>
> [DENNETT:] Right. There seems to be a pinkish glowing ring.
>
> [OTTO:] So let's talk about *that* ring.
>
> [DENNETT:] Which one?
>
> [OTTO:] The one that *seems to be*.
>
> [DENNETT:] There is no such thing as a pink ring that merely seems to be.
>
> [OTTO:] Look, I don't just *say* that there seems to be a pinkish glowing ring; *there really does seem* to be a pinkish glowing ring!

[DENNETT:] I hasten to agree.... You really mean it when you say there seems to be a pinkish glowing ring.

[OTTO:] Look. I don't just mean it. I don't just *think* there seems to be a pinkish glowing ring; *there really* seems to be a pinkish glowing ring!

[DENNETT:] Now you've done it. You've fallen in a trap, along with a lot of others. You seem to think there's a difference between thinking (judging, deciding, being of the firm opinion that) something seems pink to you and something *really seeming* pink to you. But there is no difference. There is no such phenomenon as really seeming—over and above the phenomenon of judging in one way or another that something is the case.

In other words, demands for an explanation of qualia can't be met. There are no such things.

Aaron Sloman disagrees. He acknowledges the real existence of qualia. But he does so in an unusual way: he analyses them as aspects of the multi-dimensional virtual machines that we call minds (see the following section).

Qualia, he says, are internal computational states. They can have causal effects on behaviour (e.g. involuntary facial expressions) and/or on other aspects of the mind's information processing. They can exist only in virtual machines of significant structural complexity (he outlines the types of reflexive computational resources that are required). They can be accessed only by some other parts of the particular virtual machine concerned, and don't necessarily have any behavioural expression. (Hence their *privacy*.) Moreover, they cannot always be described—by higher, self-monitoring, levels of the mind—in verbal terms. (Hence their *ineffability*.)

This doesn't mean that Sloman identifies qualia with brain processes (as Churchland does). For computational states are aspects of *virtual* machines: they can't be defined in the language of physical descriptions. But they can exist, and have causal effects, only when implemented in some underlying physical mechanism.

What of the Turing Test? Both Dennett's and Sloman's analyses imply (and Dennett explicitly argues) that zombies are impossible. That's because, for them, the concept of *zombie* is incoherent. Given the appropriate behaviour and/or virtual machine, consciousness—for Sloman, even including qualia—is guaranteed. The Turing Test is therefore saved from the objection that it might be 'passed' by a zombie.

And what of the hypothetical AGI? If Dennett is right, this would have all the consciousness that we do—which *would not* include qualia. If Sloman is right, it would have phenomenal consciousness in just the same sense that we do.

Virtual machines and the mind–body problem

Hilary Putnam's 1960s 'functionalism' used the notion of Turing machines, and the (then novel) software/hardware distinction, to argue—in effect—that *the mind is what the brain does.*

The (Cartesian) metaphysical divide between two utterly different substances gave way to a conceptual divide between levels of description. The *program versus computer* analogy allowed that 'mind' and 'body' are indeed very different. But it was fully compatible with materialism. (Whether it could encompass qualia was, and still is, hotly disputed.)

Although several intriguing AI programs existed by 1960 (see Chapter 1), the functionalist philosophers rarely considered specific examples. They focused on general principles, such as Turing computation. Only with the mid-1980s rise of PDP (see Chapter 4) did many philosophers consider how AI systems actually work. Even then, very few asked *just what* computational functions might make reasoning, or creativity (for instance), possible.

The best way of understanding such matters is to borrow the computer scientist's concept of virtual machines. Instead of saying that *the mind is what the brain does*, one should say (following Sloman) that *the mind is the virtual machine—or, rather, the integrated set of many different virtual machines—that is/are implemented in the brain.* (The mind-as-virtual-machine position has one highly counter-intuitive implication, however: see the section 'Is neuroprotein essential'.)

As explained in Chapter 1, virtual machines are real, and have real causal effects: no metaphysically mysterious mind–body interactions there. So the *philosophical* significance of LIDA, for instance, is that it specifies an organized set of virtual machines that shows how the diverse aspects of (functional) consciousness are possible.

The virtual-machine approach amends a core aspect of functionalism: the Physical Symbol System (PSS) hypothesis. In the 1970s, Allen Newell and Herbert Simon defined a PSS as 'a set of entities, called symbols, which are physical patterns that can occur as components of another type of entity called an expression (or symbol structure).... [Within] a symbol structure,... instances (or tokens) of symbols [are] related in some physical way (such as one token being next to another)'. Processes exist, they said, for creating and modifying symbol structures—namely, the processes defined by symbolic AI. And, they added, 'A PSS has the necessary and sufficient means for general intelligent action'. In other words, the mind–brain is a PSS.

From the mind-as-virtual-machine perspective, they should have called it the *Physically Implemented* Symbol System hypothesis (let's not express this as an acronym), as symbols are contents of virtual machines, not physical machines.

This implies that neural tissue isn't *necessary* for intelligence unless it's the only material substrate capable of implementing the virtual machines concerned.

The PSS hypothesis (and most early AI) assumed that a *representation*, or physical symbol, is a clearly isolatable and precisely locatable feature of the machine/brain. Connectionism would offer a very different account of representations (see Chapter 4). It saw them in terms of entire networks of cells, not clearly locatable neurons. And it saw concepts in terms of partially conflicting constraints, not strict logical definitions. This was highly attractive to philosophers familiar with Ludwig Wittgenstein's account of 'family resemblances'.

Later, workers in situated robotics denied that the brain contains representations at all (see Chapter 5). This position was accepted by some philosophers, but David Kirsh, for example, argued that compositional representations (and symbolic computation) are needed for all behaviour that involves concepts—including logic, language, and deliberative action.

Meaning and understanding

According to Newell and Simon, any PSS carrying out the right computations *really is* intelligent. It has 'the *necessary and sufficient* means for intelligent action'. The philosopher John Searle called this claim 'strong AI'. ('Weak AI' held merely that AI models can help psychologists to formulate coherent theories.)

He argued that strong AI was mistaken. Symbolic computation may go on in our heads (though he doubted it), but *that alone* cannot provide intelligence. More accurately, it cannot provide 'intentionality'—the philosophers' technical term for meaning, or understanding.

Searle relied on a thought experiment that's still controversial today: *Searle is in a windowless room, with a slot through which paper slips carrying 'squiggles' and 'squoggles' are passed in. There is a box of slips carrying similar doodles, and a rule book saying that if a squiggle is passed in then Searle should pass a blingle-blungle out, or perhaps go through a long sequence of doodle pairings before passing some slip out. Unbeknownst to Searle, the doodles are Chinese writing; the rule book is a Chinese NLP program; and the Chinese people outside the room are using him to answer their questions. However, Searle entered the room unable to understand Chinese, and he still won't understand it when he leaves.* Conclusion: *Formal computation alone (which is what Searle-in-the-room is doing) can't generate intentionality. So strong AI is mistaken, and genuine understanding in AI programs is impossible.* (This argument, the so-called 'Chinese Room', was originally aimed at symbolic AI, but was later generalized to apply to connectionism and robotics.)

Searle's claim, here, was that the 'meanings' attributed to AI programs come entirely from human users/programmers. They're arbitrary with respect to the program itself, which is semantically empty. Being 'all syntax and no semantics', the same program might be equally interpretable as a tax reckoner or as choreography.

Sometimes, that's true. But remember Franklin's claim that LIDA models *grounded*, even *embodied*, cognition, by means of structured couplings between senses, actuators, and environment. Remember, too, the control circuit that evolved as a robot's orientation detector (see Chapter 5). To call this an 'orientation detector' is *not* arbitrary. Its very existence depends on its evolution as an orientation detector, helpful in achieving the robot's goal.

The latter example is relevant not least because some philosophers see evolution as the source of intentionality. Ruth Millikan, for instance, argues that thought and language are *biological* phenomena, whose meanings depend on our evolutionary history. If that is right, then no non-evolutionary AGI could have real understanding.

Other scientifically minded philosophers (like Newell and Simon themselves) define intentionality in causal terms. But they have difficulty accounting for non-veridical statements: if someone claims to see a cow, but there's no cow there to cause the words, how can they *mean* cow?

In summary, no theory of intentionality satisfies all philosophers. Since genuine intelligence involves understanding, that's another reason why *no one knows* whether our hypothetical AGI would really be intelligent.

Is neuroprotein essential?

Part of Searle's reason for rejecting strong AI was that computers aren't made of neuroprotein. Intentionality, he said, is caused by neuroprotein much as photosynthesis is caused by chlorophyll. Neuroprotein may not be the only substance in the universe that can support intentionality and consciousness. But metal and silicon, he said, obviously can't.

That's a step too far. Admittedly, it's highly counter-intuitive to suggest that tin-can computers could really experience blueness or pain, or really understand language. But qualia being caused *by neuroprotein* is no less counter-intuitive, no less philosophically problematic. (So, something that's counter-intuitive may nevertheless be true.)

If one accepts Sloman's virtual-machine analysis of qualia, this particular difficulty vanishes. However, the *overall* mind-as-virtual-machine account brings another similar difficulty. If a mind-qualifying virtual machine were implemented in AI hardware, then *that very mind* would exist in the machine—or perhaps in several machines. So mind-as-virtual-machine implies the possibility, in principle, of (multiply cloned) personal immortality *in computers*. For most people (however, see Chapter 7), that's no less counter-intuitive than computers supporting qualia.

If neuroprotein actually is the only substance capable of supporting human-scale virtual machines, we can reject the 'cloned immortality' suggestion. But is it? We don't know.

Perhaps there are special, maybe highly abstract, properties that neuroprotein has which make it capable of implementing the wide range of computations carried out by minds. For instance, it must be able to construct (fairly rapidly) molecules that are stable (and storeable) and yet flexible, too. It must be able to form structures, and connections between structures, that have electrochemical properties enabling them to pass information between them. Possibly, other substances, on other planets, could do these things too.

Not just brain, but body too

Some philosophers of mind argue that the brain receives too much attention. The whole body, they say, is the better focus.

Their position often draws on continental phenomenology, which stresses the human 'form of life'. This covers both meaningful consciousness (including human 'interests', which ground our sense of *relevance*) and embodiment.

To be embodied is to be a living body situated in, and actively engaging with, a dynamical environment. The environment—and the engagement—is both physical and sociocultural. The key psychological properties aren't reasoning or thought, but adaptation and communication.

Philosophers of embodiment have little time for symbolic AI, seeing it as overly cerebral. Only cybernetics-based approaches are favoured (see Chapters 1 and 5). And since, on this view, genuine intelligence is body-based, no on-screen AGI could *really* be intelligent. Even if the on-screen system is an autonomous agent structurally coupled to a physical environment, it wouldn't (*pace* Franklin) count as *embodied*.

What about robots? After all, robots are physical beings grounded in, and adapting to, the real world. Indeed, *situated* robotics is sometimes commended by these philosophers. But do robots have *bodies*? Or *interests*? Or *forms of life*? Are they *alive* at all?

Phenomenologists would say 'Certainly not!' They might cite Wittgenstein's famous remark: 'If a lion could talk, we would not understand him'. The lion's form of life is so different from ours that communication would be near-impossible. Granted, there's enough overlap between a lion's psychology and ours (e.g. hunger, fear, fatigue, etc.) that some minimal understanding—and empathy—might be feasible. But even that wouldn't be available when 'communicating' with a robot. (That's why research on computer companions is so worrying: see Chapters 3 and 7.)

Moral community

Would we—should we?—accept a human-level AGI as a member of our moral community? If we did, this would have significant practical consequences. For it would affect human–computer interaction in three ways.

First, the AGI would receive our moral concern—as animals do. We would respect its interests, up to a point. If it asked someone to interrupt their rest or crossword puzzle to help it achieve a 'high-priority' goal, they'd do so. (Have you never got up out of your armchair to walk the dog, or to let a ladybird out into the garden?) The more we judged that its interests *mattered* to it, the more we'd feel obliged to respect them. However, that judgement would depend largely on whether we attributed phenomenal consciousness (including felt emotions) to the AGI.

Second, we would regard its actions as morally evaluable. Today's killer drones aren't morally responsible (unlike their users/designers: see Chapter 7). But perhaps a *truly* intelligent AGI would be? Presumably, its decisions could be affected by our reactions to them: by our praise or blame. If not, there's no *community*. It could learn to be 'moral' much as an infant (or a dog) can learn to be well behaved, or an older child to be considerate. (Consideration requires the development of what cognitive psychologists call 'Theory of Mind', which interprets people's behaviour in terms of agency, intention, and belief.) Even punishment might be justified, on instrumental grounds.

And third, we'd make it the target of argument and persuasion about moral decisions. It might even offer moral advice to people. For us to engage seriously in such conversations, we'd need to be confident that (besides having human-level intelligence) it was amenable to specifically *moral* considerations. But just what does that mean? Ethicists disagree profoundly not only about the content of morality but also about its philosophical basis.

The more one considers the implications of 'moral community', the more problematic the notion of admitting AGIs seems to be. Indeed, most people have a strong intuition that the very suggestion is absurd.

Morality, freedom, and self

That intuition arises largely because the concept of moral responsibility is intimately linked to others—conscious agency, freedom, and self—that contribute to our notion of *humanity* as such.

Conscious deliberation makes our choices more morally accountable (although unconsidered actions can be criticized, too). Moral praise or blame is attributed to the agent, or self, concerned. And actions done under strong constraints are less open to blame than those made freely.

These concepts are hugely controversial *even when applied to people.* Applying them to machines seems inappropriate—not least due to the implications for human–computer interactions cited in the previous section. Nevertheless, taking the 'mind-as-virtual-machine' approach to *human* minds can help us to understand these phenomena *in our own case.*

AI-influenced philosophers analyse freedom in terms of certain sorts of cognitive-motivational complexity. They point out that people are clearly 'free' in ways that crickets, for instance, aren't. Female crickets find their mates by a hardwired reflex response (see Chapter 5). But a woman seeking a mate has many strategies available. She also has many other motives besides mating—not all of which can be satisfied simultaneously. She manages, nevertheless—thanks to computational resources (a.k.a. intelligence) that crickets lack.

These resources, organized by functional consciousness, include perceptual learning; anticipatory planning; default assignment; preference ranking; counterfactual reasoning; and emotionally guided action scheduling. Indeed, in his book *Elbow Room* Dennett uses such concepts—and a host of telling examples—to explain human freedom. So AI helps us to understand how our own free choice is possible.

Determinism/indeterminism is largely a red herring. There is some element of indeterminism in human action, but this can't occur at the point of decision because that would undermine moral responsibility. It could, however, affect the considerations that arise during deliberation. The agent may or may not think of x, or be reminded of y—where x and y include both facts and moral values. For instance, someone's choice of a birthday present may be influenced by their accidentally noticing something that reminds them that the potential recipient likes purple, or supports animals' rights.

All the computational resources just listed would be available to a human-level AGI. So, unless free choice must also involve *phenomenal* consciousness (and if one rejects computational analyses of that), it seems that our imaginary AGI would have freedom. If we could make sense of the AGI's having various motives that *mattered* to it, then distinctions could even be made between its choosing 'freely' or 'under constraint'. However, that 'if' is a very big one.

As for the self, AI researchers stress the role of *recursive* computation, in which a process can operate upon itself. Many traditional philosophical puzzles concerning self-knowledge (and self-deception) can be dissolved by this AI-familiar idea.

But what is 'self-knowledge' knowledge *of*? Some philosophers deny the reality of the self, but AI-influenced thinkers don't. They see it as a specific type of virtual machine.

For them, the self is an enduring computational structure that organizes and rationalizes the agent's actions—especially their carefully considered voluntary actions. (LIDA's author, for instance, describes it as 'the enduring context of experience, that organizes and stabilizes experiences across many different local contexts'.) It isn't present in the newborn baby, but is a lifelong construction—to some extent amenable to deliberate self-moulding. And its multi-dimensionality allows for considerable variation, generating recognizably *individual* agency, and *personal* idiosyncrasy.

That's possible because the agent's Theory of Mind (which initially interprets the behaviour of others) is applied, reflexively, to one's own thoughts and actions. It makes sense of them in terms of prioritized motives, intentions, and goals. These, in turn, are organized by enduring individual preferences, personal relationships, and moral/political values. This computational architecture allows for the construction of both *self-image* (representing the sort of person one believes one is) and *ideal self-image* (the sort of person one would like to be), and for actions and emotions grounded in the differences between the two.

Dennett (strongly influenced by Minsky) calls the self 'the center of narrative gravity': a structure (virtual machine) that, in telling the story of one's own life, both generates and seeks to explain one's actions—especially one's relationships with other people. This leaves room, of course, for self-deception and self-invisibility of numerous kinds.

Similarly, Douglas Hofstadter describes selves as abstract self-referential patterns that arise from and causally loop back into the meaningless base of neural activity. These patterns (virtual machines) aren't superficial aspects of the person. To the contrary, for the self to exist *just is* for that pattern to be instantiated.

In sum: deciding to credit AGIs with *real* human-level intelligence—involving morality, freedom, and self—would be a big step, with significant practical implications. Those whose intuition rejects the whole idea as fundamentally mistaken may well be correct. Unfortunately, their intuition can't be buttressed by non-controversial philosophical arguments. There's no consensus on these matters, so there are no easy answers.

Mind and life

All the minds we know about are found in living organisms. Many people—including the cyberneticians (see Chapters 1 and 5)—believe this *must* be so. That is, they assume that mind necessarily presupposes life.

Professional philosophers sometimes state this explicitly, but rarely argue for it. Putnam, for instance, said it's an 'undoubted fact' that if a robot isn't alive then it can't be conscious. But he gave no scientific reasons, relying instead on 'the semantical rules of our language'. Even the few people—such as the environmentalist philosopher Hans Jonas, and recently the physicist Karl Friston, via his broadly cybernetic 'free-energy principle'—who have defended the assumption at length haven't proved it beyond doubt.

Let's assume, however, that this common belief is true. If so, then real intelligence can be achieved by AI only if real life is achieved, too. We must ask, then, whether 'strong A-Life' (life in cyberspace) is possible.

There's no universally accepted definition of life. But nine features are usually mentioned: self-organization, autonomy, emergence, development, adaptation, responsiveness, reproduction, evolution, and metabolism. The first eight can be understood in information-processing terms, so could in principle be instantiated by AI/A-Life. Self-organization, for instance—which, broadly understood, includes all the others—has been achieved in various ways (see Chapters 4 and 5).

But metabolism is different. It can be *modelled* by computers, but not *instantiated* by them. Neither self-assembling robots nor virtual (on-screen) A-Life can actually metabolize. Metabolism is the use of biochemical substances and energy exchanges to assemble, and maintain, the organism. So it's irreducibly physical. Defenders of strong A-Life point out that computers use energy, and that some robots have *individual* energy stores, needing regular replenishment. But that's a far cry from the flexible use of interlocking biochemical cycles to build the organism's bodily fabric.

So if metabolism is necessary for life, then strong A-Life is impossible. And if life is necessary for mind, then strong AI is impossible too. No matter how impressive the performance of some future AGI, it wouldn't have intelligence, *really*.

The great philosophical divide

'Analytic' philosophers, and AI researchers too, take it for granted that *some* scientific psychology is possible. Indeed, that position has been assumed throughout this book—including this chapter.

Phenomenologists, however, reject that assumption. They argue that all our scientific concepts *arise from* meaningful consciousness, so can't be used to *explain* it. Before he died in 2016, Putnam himself had accepted that position. They even claim that it's nonsensical to posit a real world existing independently of human thought, whose objective properties science may discover.

So the lack of consensus about the nature of mind/intelligence is even deeper than I've indicated so far.

There's no knock-down argument against the phenomenologists' view—nor for it, either. For there's no common ground from which to mount one. Each side defends itself and criticizes the other, but using arguments whose key terms aren't mutually agreed. Analytic and phenomenological philosophy give fundamentally different interpretations even of basic concepts like *reason* and *truth*. (The AI scientist Brian Cantwell Smith has offered an ambitious metaphysics of *computation*, *intentionality*, and *objects* that aims to respect the insights of both sides; unfortunately, his intriguing argument is unpersuasive.)

This dispute is unresolved—and perhaps irresolvable. To some people, the phenomenologists' position is 'obviously' right. To others, it's 'obviously' absurd. That's yet another reason why *no one knows*, for sure, whether an AGI could really be intelligent.

Chapter 7
The Singularity

AI's future has been hyped since its inception. Overly enthusiastic predictions from (some) AI professionals have excited, and sometimes terrified, journalists and cultural commentators. Today, the prime example is the Singularity: the proposed point in time at which machines become more intelligent than humans.

First, it's said, AI will reach human-level intelligence. (It's tacitly assumed that this would be *real* intelligence: see Chapter 6.) Soon afterwards, AGI will morph into ASI—'S' for Superhuman. For the systems will be intelligent enough to copy themselves, and so outnumber us—and to improve themselves, and so out-think us. Most important problems and decisions will then be addressed by computers.

This notion is hugely contentious. People disagree about whether it could happen, whether it will happen, when it might happen, and whether it would be a good or bad thing.

Singularity believers (S-believers) argue that AI advances make the Singularity inevitable. Some welcome this. They foresee humanity's problems solved. War, disease, hunger, boredom, even personal death... all banished. Others predict the end of humanity—or anyway, of civilized life as we know it. Stephen Hawking (alongside Stuart Russell, co-author of AI's leading textbook) made worldwide waves in May 2014 by saying that to ignore the threat of AI would be 'potentially our worst mistake ever'.

By contrast, the Singularity sceptics (S-sceptics) don't expect the Singularity to happen—and certainly not in the foreseeable future. They allow that AI provides plenty to worry about. But they don't see an existential threat.

Prophets of the Singularity

The idea of an AGI-to-ASI transition has recently become a media commonplace, but it originated in the mid-20th century. The key initiators were 'Jack' Good (a fellow codebreaker of Alan Turing's at Bletchley Park), Vernor Vinge, and Ray Kurzweil. (Turing himself had expected 'the machines to take control', but didn't elaborate.)

In 1965, Good predicted an ultraintelligent machine, which would 'far surpass all the intellectual activities of any man however clever'. Since it could design even better machines, it would 'unquestionably [lead to] an intelligence explosion'. At that time, Good was cautiously optimistic: 'The first ultraintelligent machine is the last invention that man need ever make—*provided that the machine is docile enough to tell us how to keep it under control*'. Later, however, he argued that ultraintelligent machines would destroy us.

A quarter-century later, Vinge popularized the term 'Singularity' (initiated in this context by John von Neumann in 1958). He predicted 'The Coming Technological Singularity'—at which all predictions will break down (compare the event horizon of a black hole).

The Singularity itself, he allowed, *can* be foreseen: indeed, it's inevitable. But among the many (unknowable) consequences might be the destruction of civilization, and even of humankind. We're heading for 'a throwing away of all the previous rules, perhaps in the blink of an eye, an exponential runaway beyond any hope of control'. Even if *every* government realized the danger and tried to prevent it, he said, they couldn't.

The pessimism of Vinge and (eventually) Good is countered by Kurzweil. He offers not only breath-taking optimism, but also dates.

His book, tellingly titled *The Singularity is Near*, suggests that AGI will be achieved by 2030—and that by 2045, ASI (combined with nanotechnology and biotechnology) will have defeated war, disease, poverty, and personal death. It will also have engendered 'an explosion of art, science, and other forms of knowledge that... will make life truly meaningful'. By mid-century, too, we'll be living in immersive virtual realities hugely more rich and satisfying than the real world. For Kurzweil, '*The Singularity*' really is singular, and '*Near*' really does mean near.

(This hyper-optimism is sometimes tempered. Kurzweil lists many existential risks, largely from AI-aided biotechnology. Regarding AI itself, he says: 'Intelligence is inherently impossible to control... It is infeasible today to devise strategies that will absolutely ensure that future AI embodies human ethics and values'.)

Kurzweil's argument relies on 'Moore's Law', the observation—by Gordon Moore, who founded Intel—that the computer power available for one dollar doubles every year. (The laws of physics will conquer Moore's Law eventually, but not in the foreseeable future.) As Kurzweil points out, *any* exponential increase is highly counter-intuitive. Here, he says, it implies AI advance at an unimaginable rate. So, like Vinge, he insists that expectations built on past experience are near-worthless.

Competing predictions

Despite being declared near-worthless, post-Singularity forecasts are often made nevertheless. A host of mind-boggling examples are found in the literature, of which only a few can be mentioned here.

S-believers fall into two camps: pessimists (following Vinge) and optimists (following Kurzweil). They mostly agree that AGI-to-ASI will happen well before the end of this century. But they disagree on just how dangerous the ASI will be.

For example, some foresee evil robots doing all in their power to thwart human hopes and lives (a common trope of science fiction and Hollywood movies). The idea that we could 'pull the plug', if necessary, is specifically denied. The ASIs, we're told, would be canny enough to make this impossible.

Others argue that ASIs will have no malicious intent *but will be hugely dangerous anyway*. We wouldn't build hatred of humans into them, and there's no reason why they should develop it for themselves. Rather, they will be indifferent to us, much as we are to most non-human species. But their indifference, if our interests conflict with their own goals, could be our downfall: *Homo sapiens* as dodo. In Nick Bostrom's widely quoted thought experiment, an ASI intent on making paper clips would scavenge the atoms in human bodies to manufacture them.

Or again, consider a general strategy sometimes suggested for guarding against Singularity threats: *containment*. Here, an ASI is prevented from directly acting on the world, although it can directly perceive the world. It's used only to answer our questions (what Bostrom calls an 'Oracle'). However, the world includes the Internet, and ASIs might cause changes indirectly by contributing content—facts, falsehoods, computer viruses...—to the Internet.

Another form of Singularity pessimism predicts that the machines will get us to do their dirty work for them, even if this goes against humanity's interests. This view scorns the idea that we could 'contain' ASI systems by cutting them off from the world. A super-intelligent machine, it's said, could use bribery or threats to persuade one of the few humans to which it is sometimes connected to do things that it's unable to do directly.

That particular worry assumes that the ASI will have learned enough about human psychology to know what bribes or threats are likely to work, and maybe also which individuals are most likely to be vulnerable to a certain type of persuasion. To the objection that this assumption is incredible, the reply would be that crude financial bribes, or murderous threats, would work with almost anyone—so the ASI wouldn't need psychological insight rivalling that of Henry James. Nor would it need to *understand*, in human terms, what *persuasion*, *bribery*, and *threat* actually are. It would merely need to know that inputting certain NLP texts to a human being is likely to influence their behaviour in broadly predictable ways.

Some of the optimistic forecasts are even more challenging. Perhaps the most arresting are Kurzweil's predictions of living in a virtual world and of the elimination of personal death. Bodily death, although much delayed (by ASI-aided bioscience), would continue. But death's sting could be drawn by downloading the personalities and memories of individual people into computers.

This philosophically problematic assumption, that a person could exist either in silicon or in neuroprotein (see Chapter 6), is reflected in the subtitle of his 2005 book: *When Humans Transcend Biology*. Kurzweil was expressing his 'Singularitarian' vision—also called transhumanism, or posthumanism—of a world containing partly, or even wholly, non-biological people.

These transhumanist 'cyborgs', it's claimed, will have various computerized implants directly connected to their brains, and prostheses for limbs and/or sense organs. Blindness and deafness will be banished, because visual and auditory signals will be interpreted through the sense of touch. Not least, rational cognition (as well as moods) will be enhanced by specially designed drugs.

Early versions of such assistive technologies are already with us. If they proliferate as Kurzweil suggests, our concept of humanity will be profoundly changed. Instead of seeing prostheses as useful add-ons to human bodies, they will be seen as (trans)human body parts. Widely used psychotropic drugs will be listed alongside natural substances like dopamine in accounts of 'the brain'. And the superior intelligence, strength, or beauty of genetically engineered individuals will be regarded as 'natural' features. Political views on egalitarianism and democracy will be challenged. A new sub-species (or species?) may even develop, from human ancestors wealthy enough to exploit these possibilities.

In short, biological evolution is expected to be replaced by technological evolution. Kurzweil sees the Singularity as 'the culmination of the merger of our biological thinking and existence with our technology, resulting in a world [in which] there will be no distinction...between human and machine or between physical and virtual reality'. (You can be forgiven for feeling that you need to take a very deep breath.)

Transhumanism is an extreme example of how AI may change ideas about human nature. A less extreme philosophy that assimilates technology into the very concept of *mind* is 'the extended mind', seeing the mind as spread out over the world to include cognitive processes that rely upon it. Although the notion of the extended mind has been widely influential, transhumanism has not. It has been enthusiastically endorsed by some philosophers, cultural commentators, and artists. However, not all S-believers accept it.

Scepticism defended

In my judgement, the S-sceptics are right. The discussion of mind-as-virtual-machine in Chapter 6 implied that there is no obstacle *in principle* to human-level AI intelligence (possibly excepting phenomenal consciousness). The question here is whether this is likely *in practice*.

Besides the intuitive implausibility of many post-Singularity predictions, and the near-absurdity (in my opinion) of transhumanist philosophy, the S-sceptics have other arguments on their side.

AI is less promising than many people assume. Chapters 2 through 5 mentioned countless things that current AI *cannot* do. Many require a human sense of *relevance* (and tacitly assume completion of the semantic web: see Chapter 2). Moreover, AI has focused on intellectual rationality while ignoring social/emotional intelligence—never mind wisdom. An AGI that could interact fully with our world would need those capacities, too. Add the prodigious richness of human minds, and the need for good psychological/computational theories about how they work, and the prospects for human-level AGI look dim.

Even if it were practically feasible, it's doubtful whether the necessary funding would materialize. Governments are currently putting enormous resources into brain emulation (see the following section), but the money required for building artificial human minds would be much greater still.

Thanks to Moore's Law, further AI advances can certainly be expected. But increases in computer power, and in data availability (given cloud storage, and 24/7 sensors throughout the Internet of

Things), won't guarantee human-like AI. That's bad news for S-believers, because ASI needs AGI first.

S-believers ignore the limitations of current AI. They simply don't care, because they have a trump card: the notion that exponential technological advance is rewriting all the rule books. This licenses them to make predictions at will. They occasionally allow that 'by-end-of-the-century' predictions may be unrealistic. However, they insist that 'never' is a long time.

Never is indeed a long time. So the S-sceptics, myself included, may be wrong. They have no knock-down argument—especially if they allow AGI's possibility *in principle* (as I do). They may even be persuaded that the Singularity, albeit hugely delayed, will happen eventually.

Nevertheless, careful consideration of state-of-the-art AI gives good reason to back the sceptics' hypothesis (or their *bet*, if you prefer), rather than the wild speculations of the S-believers.

Whole-brain emulation

S-believers predict exponential technological advance in AI, biotechnology, and nanotechnology—and in cooperation between them. Indeed, it's already happening. Analyses of Big Data are being used to advance genetic engineering and drug development, and many other scientifically based projects too (Ada Lovelace vindicated: see Chapter 1). Likewise, AI and neuroscience are being combined in whole-brain emulation (WBE).

The aim of WBE is to mimic a real brain by simulating its individual components (neurons), along with their connections and information-processing capabilities. The hope is that the scientific knowledge acquired will have many applications, including treatments for mental pathologies ranging from Alzheimer's to schizophrenia.

This reverse engineering will require neuromorphic computing, which models sub-cellular processes such as the passage of ions through the cell membrane (see Chapter 4).

Neuromorphic computing depends on knowledge about the anatomy and physiology of the various types of neuron. But WBE will also need detailed evidence about specific neuronal connections and functionality, including timing. Much of this will require improved brain scanning, with miniaturized neuroprobes monitoring individual neurons continuously.

Various WBE projects are now underway, often compared by their sponsors to the Human Genome Project or the race to the moon. For instance, in 2013 the European Union announced the Human Brain Project, costed at £1 billion. Later that year, US President Barack Obama heralded BRAIN, a ten-year project funded by $3 billion from the US government (plus a significant amount of private money). It aims first to generate a dynamic map of the connectivity of the mouse brain, and then to emulate the human case.

Earlier attempts at *partial* brain emulation were also government-funded. In 2005, Switzerland sponsored the *Blue Brain* project—initially to simulate a rat's cortical column, but with the long-term aim of modelling the million columns in the human neocortex. In 2008, DARPA provided nearly $40 million for SyNAPSE (Systems of Neuromorphic and Plastic Scalable Electronics); by 2014—and with a further $40 million—this was using chips carrying 5.4 billion transistors, each holding one million units (neurons) and 256 million synapses. And Germany and Japan are collaborating in using NEST (NEural Simulation Technology) to develop the K computer; by 2012, this was taking forty minutes to simulate one second of 1 per cent of real-brain activity, involving 1.73 billion 'neurons' and 10.4 trillion 'synapses'.

Because it's so expensive, mammalian WBE is rare. But countless attempts to map much smaller brains are going on around the world (at my own university, they're focused on honey bees). These may provide neuroscientific insights that can help human-scale WBE.

Given the hardware progress already achieved (e.g. SyNAPSE's chips), plus Moore's Law, Kurzweil's prediction that computers matching the raw processing power of human brains will exist by the 2020s is plausible. But his belief that they will match human intelligence by 2030 is another matter.

For it is the *virtual* machine that's crucial here (see Chapters 1 and 6). Some virtual machines can be implemented only in hugely powerful hardware. So mega-transistored computer chips may well be necessary. But just what computations will they be performing? In other words, just what virtual machines will be implemented in them? To match human (or even mouse) intelligence, these will need to be *informationally* powerful, in ways that computational psychologists don't yet fully understand.

Let's suppose—what is, I think, unlikely—that every neuron in a human brain will, eventually, be mapped. In itself, this won't tell us what they are *doing*. (The tiny nematode worm *C. elegans* has only 302 neurons, whose connections are known precisely. But we can't even identify the synapses as excitatory/inhibitory.)

For visual cortex, we already have a fairly detailed mapping between neuroanatomy and psychological function. But that's not so for neocortex in general. In particular, we don't know much about what it is that the frontal cortex is doing—that is, what virtual machines are implemented in it. That question isn't prominent in large-scale WBE. The Human Brain Project, for instance, has adopted a firmly bottom-up approach: look at the anatomy and biochemistry, and mimic it. Top-down questions, about the psychological functions that the brain may be supporting, are sidelined (very few cognitive neuroscientists are involved). Even if the anatomical modelling were fully achieved, and the chemical messaging carefully monitored, those top-down questions wouldn't be answered.

Answers would require a wide variety of computational concepts. Moreover, one key topic is the computational architecture of the mind (or mind–brain) *as a whole.* We saw in Chapter 3 that action planning in multi-motive creatures requires complex scheduling mechanisms—such as those provided by emotions. And the discussion of LIDA in Chapter 6 indicated the huge complexity of cortical processing. Even the mundane activity of eating with a knife and fork requires that many virtual machines be integrated—some dealing with physical objects (muscles, fingers, utensils, various kinds of sensors), others with intentions, plans, expectations, desires, social conventions, and preferences. To understand how that activity is possible, we need not only neuroscientific data about the brain, but also detailed computational theories about the psychological processes involved.

In short, considered as a route to understanding human intelligence, bottom-up WBE is likely to fail. It may teach us a lot about the brain. And it may help AI scientists to develop further practical applications. But the notion that mid-century WBE will have explained human intelligence is an illusion.

What we should be worrying about

If the S-sceptics are right, and there will be no Singularity, it doesn't follow that there's nothing to worry about. AI already raises matters of concern. Future advances will surely raise more, so anxiety about the long-term safety of AI isn't wholly out of place. More to the point, we need to pay attention to its short-term influences, too.

Some worries are very general. For example, any technology can be used for good or ill. Malicious people will use any available tools—and sometimes fund the development of new ones—to do malicious things. (CYC, for instance, might be useful for wrongdoers: its developers are already thinking about how to limit access to the full system, when released—see Chapter 2.) So we must be very careful about what we invent.

As Stuart Russell points out, this means more than merely being careful about our *aims*. If there are ten parameters relevant to a problem, and statistically optimizing machine learning (see Chapter 2) considers only six, then the other four can—and probably will—be pushed to extremes. Hence we also need to be vigilant about *what sorts of data* are being used.

That general worry concerns the frame problem (see Chapter 2). Like the fisherman in the fairy story, whose wish for his soldier-son to come home was granted by his being brought back in a coffin, we could be nastily surprised by powerful AI systems lacking our understanding of *relevance*.

For example, when a Cold War early-warning system (on 5 October 1960) recommended a defensive strike on the USSR, disaster was averted only by the operators' sense of relevance—both political and humanitarian. They judged that the Soviets at the UN hadn't been especially obstreperous recently, and they feared the horrendous consequences of a nuclear attack. So, violating protocols, they ignored the automated warning. Several other nuclear near-misses have happened; some, recently. Usually, escalation was prevented only by people's common sense.

Moreover, human error is always possible. Sometimes, it's understandable. (The Three Mile Island emergency was made worse by humans overriding the computer, but the physical conditions they were facing were *highly* unusual.) But it can be breathtakingly unexpected. The Cold War alert mentioned in the previous paragraph happened because someone had forgotten leap years when programming the calendar—so the moon was in the 'wrong' place. All the more reason, then, for testing and (if possible) proving the reliability of AI programs before they're used.

Other worries are more specific. Some should be troubling us today.

One major threat is technological unemployment. Many manual and low-level clerical jobs have disappeared. Others will follow (although manual jobs requiring dexterity and adaptability won't vanish). Most of the lifting, fetching, and carrying in a warehouse can now be done by robots. And driverless vehicles will mean jobless people.

Middle-managerial positions are at risk also. Many professionals are already using AI systems as aids. It won't be long before jobs (in law and accountancy, for instance) involving time-consuming research into regulations and precedents can be largely taken over by AI. More demanding tasks, including many in medicine and science, will fairly soon be affected too. Jobs will be down-skilled, even if they're not lost. And professional training will suffer: how will the youngsters learn to make sensible judgements?

While some legal jobs will be redundant, lawyers will also gain from AI, because a host of legal traps lie in wait. If something goes wrong, who should be responsible: programmer, wholesaler, retailer, or user? And might a human professional sometimes be sued for *not* using an AI system? If the system had been shown (whether mathematically or empirically) to be highly reliable, such litigation would be very likely.

New types of jobs will doubtless appear. But whether these will be equivalent in terms of numbers, educational accessibility, and/or breadwinning power (as happened after the Industrial Revolution) is doubtful. Serious sociopolitical challenges lie ahead.

'Service' positions are less threatened. But even they are endangered. In an ideal world, the opportunity for multiplying, and upgrading, currently undervalued person-to-person activities would be grasped with enthusiasm. However, that's not guaranteed.

For instance, education is being opened up to personal and/or Internet-based AI aids—including MOOCs (Massive Open Online Courses) offering lectures by academic stars—that down-skill the jobs of many human teachers. Computer psychotherapists are already available, at much less expense than human therapists. (Some are surprisingly helpful—for recognizing depression, for instance.) However, they're entirely unregulated. And we saw in Chapter 3 that demographic change is encouraging research in the potentially lucrative area of artificial 'carers' for the elderly, as well as 'robot nannies'.

Quite apart from any effects on unemployment, the use of empathy-free AI systems in such essentially human contexts is both practically risky and ethically dubious. Many 'computer companions' are designed for use by elderly and/or disabled people who have only minimal personal contact with the few humans they encounter. They're intended as sources not merely of aid and entertainment but also of conversation, conviviality, and emotional comfort. Even if the vulnerable person is made happier by such technology (as *Paro*-users are), their human dignity is insidiously betrayed. (Cultural differences are important here: attitudes to robots differ hugely between Japan and the West, for instance.)

The elderly users may enjoy discussing their personal memories with an artificial companion. But is this really a *discussion*? It might be a welcome reminder, triggering comforting episodes of nostalgia. However, that benefit could be provided without seducing the user into an illusion of empathy. Often, even in emotionally fraught counselling situations, what the person wants above all is an *acknowledgement* of their courage and/or suffering. But that arises from a shared understanding of the human condition. We are short-changing the individual by offering only a superficial simulacrum of sympathy.

Even if the user is suffering moderately from dementia, their 'theory' of the AI agent is likely to be much richer than the agent's model of the human. What would result, then, if the agent fails to respond as expected, and as *needed*, when the person reminisces about some distressing personal loss (of a child, perhaps)? Conventional expressions of sympathy from the companion wouldn't help—and might do more harm than good. Meanwhile, the person's distress would have been aroused with no comfort immediately available.

Another worry concerns whether the companion should sometimes be silent, or tell a white lie. Relentless truth-telling (and sudden silences) might upset the user. But tactfulness would require hugely advanced NLP plus a subtle model of human psychology.

As for robot nannies (and ignoring safety issues), overuse of AI systems with babies and infants might warp their social and/or linguistic development.

Artificial sex partners are not only being depicted in the movies (in the film *Her*, for instance). They are already being marketed. Some are capable of speech recognition, and of seductive speech and/or movement. They augment the Internet-based influences that are currently coarsening people's sexual experience (and reinforcing the sexual objectification of women). Many commentators (including some AI scientists) have written about sexual encounters with robots in terms that reveal an extraordinarily shallow concept of personal love, close to confusing it with lust, sexual obsession, and mere comfortable familiarity. However, such cautionary observations are unlikely to be effective. Given the huge profitability of pornography in general, there's scant hope of preventing future 'advances' in AI sex dolls.

Privacy is another knotty issue. It's becoming even more contentious, as powerful AI search and AI learning are let loose on data collected from personal media and home-based or wearable sensors. (Google has patented a robotic teddy bear, with camera eyes, microphone ears, and speakers in its mouth. It will be able to communicate with parents as well as child—and, willy-nilly, with unseen data collectors too.)

Cyber security has long been a problem. The more that AI enters into our world (often in very non-transparent ways), the more important it will be. One defence against an ASI takeover would be to find ways of writing algorithms that couldn't be hacked/altered (a goal of 'Friendly AI': see the next section).

Military applications, too, raise concerns. Robot minesweepers are very welcome. But robot soldiers or robot weapons? Current drones are human-instigated, but even so they may increase suffering by enlarging the *human* (not just geographical) distance between operator and target. One must hope that future drones won't be allowed to decide who/what should be a target. Even trusting them to *recognize* a (humanly chosen) target raises ethically troubling issues.

What's being done about it

None of these worries is new, although few AI workers have taken much notice of them until now.

Several AI pioneers considered the social implications at a meeting at Lake Como in 1972, but John McCarthy refused to join them, saying it was too early to speculate. A few years later, the computer scientist Joseph Weizenbaum published a book subtitled *From Judgment to Calculation*, bewailing the 'obscenity' of confusing the two—but he was dismissed contemptuously by the AI community.

There were some exceptions, of course. For instance, the first book to overview AI (written by myself, and published in 1977) included a final chapter on 'Social Significance'. And CPSR (Computer

Professionals for Social Responsibility) was founded in 1983 (partly through the efforts of SHRDLU's author Terry Winograd: see Chapter 3). But that was done primarily to warn of the unreliability of Star Wars technology—the computer scientist David Parnas even addressed the US Senate about this. As Cold War worries receded, most AI professionals seemed less concerned about their field. Only a few, such as the University of Sheffield's Noel Sharkey (a roboticist who chairs the International Committee for Robot Arms Control), plus some philosophers of AI, for example, Yale's Wendell Wallach, and Sussex's Blay Whitby, continued over the years to focus on social/ethical issues.

Now, because of both the practice and the promise of AI, the misgivings have become more pressing. Within the field (and, to some extent, outside it) social implications are receiving more attention.

Some important responses have nothing to do with the Singularity. For instance, the UN and Human Rights Watch have long advocated a treaty (not yet signed) banning fully autonomous weapons, such as target-selecting drones. And some long-established professional bodies have recently reviewed their research priorities and/or codes of conduct. But talk of the Singularity has brought additional contributors into the debate.

Many people—S-believers and S-sceptics alike—argue that even if the probability of the Singularity is extremely small, the possible consequences are so grave that we should start taking precautions now. Despite Vinge's claim that nothing can be done about the existential threat, several institutions have been founded to guard against it.

They include the UK's Centre for the Study of Existential Risk (CSER) in Cambridge and Future of Humanity Institute (FHI) in Oxford, and the USA's Future of Life Institute (FLI) in Boston and Machine Intelligence Research Institute (MIRI) in Berkeley.

These organizations are largely funded by AI philanthropists. For instance, CSER and FLI were co-founded by Jaan Tallinn, the co-developer of Skype. Both those institutions, besides communicating with AI professionals, are trying to alert policy-makers and other influential members of the public to the dangers.

The president of the American Association for AI (Eric Horwitz) organized a small panel in 2009 to discuss what precautions might be necessary to guide, or even *delay*, socially problematic AI work. This was pointedly held in Asilomar, California, where professional geneticists some years earlier had agreed a moratorium on certain genetic research. However, as a member of the group, it was my impression that not all of the participants were seriously concerned about AI's future. The ensuing report didn't get extensive media coverage.

A similarly motivated, but larger, meeting (under Chatham House rules, and with no journalists present) was convened by FLI and CSER in Puerto Rico in January 2015. The organizer, Max Tegmark, had co-signed the minatory letter with Russell and Hawking six months earlier. Unsurprisingly, then, the atmosphere was noticeably more urgent than at Asilomar. It immediately resulted in generous new funding (from the Internet millionaire Elon Musk) for research on AI safety and ethical AI—plus a cautionary open letter, signed by thousands of AI workers and widely circulated in the media.

Soon afterwards, a second open letter drafted by Tom Mitchell and several other leading researchers warned against developing autonomous weapons that would select and engage targets without human intervention. The signatories hoped to 'prevent a global AI arms race from starting'. Presented at AI's International Conference in July 2015, this was signed by nearly 3,000 AI scientists and by 17,000 people in related fields, and received extensive media coverage.

The Puerto Rico meeting also led to an open letter (in June 2015) by MIT economists Erik Brynjolfsson and Andy McAfee. This was aimed at policy-makers, entrepreneurs, and businessmen, as well as at professional economists. Warning of the potentially radical economic implications of AI, they suggested some public policy recommendations that might ameliorate—though not cancel—the risks.

January 2017 saw the second invitation-only meeting on beneficial AI. Organized, again, by Tegmark, this took place in the iconic Asilomar setting.

These AI-community efforts are persuading the transatlantic governmental funders of the importance of social/ethical issues. The USA's Department of Defense and National Science Foundation have both recently said that they are willing to fund such research. But this support isn't entirely new: 'governmental' interest has been growing for some years.

For instance, two UK Research Councils sponsored an interdisciplinary 'Robotics Retreat' in 2010, partly to draft a code of conduct for roboticists. Five 'Principles' were agreed, two of which addressed worries discussed earlier: '(1) Robots should not be designed as weapons, except for national security reasons' and '(4) Robots are manufactured artefacts: the illusion of emotions and intent should not be used to exploit vulnerable users'.

Two more laid the moral responsibility squarely on *human* shoulders: '(2) Humans, not robots, are responsible agents...' and '(5) It should be possible to find out who is [legally] responsible for any robot'. The group refrained from trying to update Isaac Asimov's 'Three Laws of Robotics' (a robot must not harm a human being, and must obey human orders and protect its own survival unless these conflict with the first Law). They insisted that any 'laws' here are to be followed by *the human designer/builder*, not the robot.

In May 2014, an academic initiative funded by the US Navy ($7.5 million for five years) was hailed across the media. This is a five-university project (Yale, Brown, Tufts, Georgetown, and Rensselaer Institute), aimed at developing 'moral competence' in robots. It involves cognitive and social psychologists and moral philosophers, as well as AI programmers and engineers.

This interdisciplinary group isn't trying to provide a list of moral algorithms (comparable to Asimov's Laws), nor to prioritize a particular meta-ethics (e.g. utilitarianism), nor even to define a set of non-competing moral values. Rather, it hopes to develop a computational system capable of moral reasoning (and moral *discussion*) in the real world. For autonomous robots will sometimes be taking deliberative decisions, not merely following instructions (still less, reacting inflexibly to 'situated' cues: see Chapter 5). If a robot is engaged in search and rescue, for instance, who should it evacuate/rescue first? Or if it's providing social companionship, when—if ever—should it avoid telling its user the truth?

The proposed system would integrate perception, motor action, NLP, reasoning (both deductive and analogical), and emotion. The latter would include emotional thinking (which can signal significant events, and also schedule conflicting goals: see Chapter 3); robotic displays of 'protest and distress', which could influence the moral decisions taken by people interacting with it; and the recognition of emotions in the humans surrounding it. And, so the official announcement declares, the robot might even 'exceed' ordinary (i.e. human) moral competence.

Given the obstacles to AGI noted in Chapters 2 and 3, plus the difficulties relating specifically to morality (see Chapter 6), one may well doubt that this task is achievable. But the project could be worthwhile nevertheless. For in considering real-world problems (like the two very different examples given earlier), it may alert us to the many dangers of using AI in morally problematic situations.

Besides these institutional efforts, an increasing number of individual AI scientists are aiming for what Eliezer Yudkowsky calls 'Friendly AI'. This is an AI that has positive effects for humanity, being both safe and useful. It would involve algorithms that are intelligible, reliable, and robust, and which fail gracefully if they fail at all. They should be transparent, predictable, and not vulnerable to manipulation by hackers—and if their reliability can be proved by logic or mathematics, as opposed to empirical testing, so much the better.

The $6 million donated by Musk at the Puerto Rico meeting led immediately to an unprecedented 'Call for Proposals' from FLI (six months later, thirty-seven projects had been funded). Aimed at experts in 'public policy, law, ethics, economics, or education and outreach' as well as in AI, it asked for: 'Research projects aimed to maximize the future societal benefit of artificial intelligence while avoiding potential hazards' and 'limited to research that explicitly focuses not on the standard goal of making AI more capable, but on making AI more robust and/or beneficial....' That welcome appeal for Friendly AI might, perhaps, have happened anyway. But the footprint of the Singularity was visible: 'Priority will be given', it said, 'to research aimed at keeping AI robust and beneficial even if it comes to greatly supersede current capabilities....'

In sum, near-apocalyptic visions of AI's future are illusory. But, partly because of them, the AI community—and policy-makers and the general public, too—are waking up to some very real dangers. Not before time.

References

NB: 'MasM', in the chapter references that follow, identifies the most relevant parts of Boden, *Mind as Machine*.

(For the analytical table of contents of MasM, see the 'Key Publications' section of my website: <www.ruskin.tv/margaretboden>.)

Chapter 1: What is artificial intelligence?

MasM chaps. 1.i.a, 3.ii–v, 4, 6.iii–iv, 10–11.

The quotations from Ada Lovelace are from: Lovelace, A. A. (1843), 'Notes by the Translator'. Reprinted in R. A. Hyman (ed.) (1989), *Science and Reform: Selected Works of Charles Babbage* (Cambridge: Cambridge University Press), 267–311.

Blake, D. V., and Uttley, A.M. (eds.) (1959), *The Mechanization of Thought Processes*, vol. 1 (London: Her Majesty's Stationery Office).

This contains several important early papers, including descriptions of Pandemonium and Perceptrons and a discussion of AI and common sense.

McCulloch, W. S., and Pitts, W. H. (1943), 'A Logical Calculus of the Ideas Immanent in Nervous Activity', *Bulletin of Mathematical Biophysics*, 5: 115–33. Reprinted in S. Papert, ed. (1965), *Embodiments of Mind* (Cambridge, MA: MIT Press), 19–39.

Feigenbaum, E. A., and Feldman, J. A., eds. (1963), *Computers and Thought* (New York: McGraw-Hill).

An influential collection of early papers on AI.

Chapter 2: General intelligence as the Holy Grail

MasM, sections. 6.iii, 7.iv, and chaps. 10, 11, 13.

Boukhtouta, A. et al. (2005), *Description and Analysis of Military Planning Systems* (Quebec: Canadian Defence and Development Technical Report).

Shows how AI planning has advanced since the early days.

Mnih, V., and D. Hassabis et al. (nineteen authors) (2015), 'Human-Level Control Through Deep Reinforcement Learning', *Nature,* 518: 529–33.

This paper by the DeepMind team describes the Atari game-player.

Silver, D., and D. Hassabis et al. (seventeen authors) (2017), 'Mastering the Game of Go Without Human Knowledge', *Nature,* 550: 354–9.

This describes the latest version of DeepMind's (2016), *AlphaGo* program (for the earlier version, see *Nature,* 529: 484–9).

The quotation from Allen Newell and Herbert Simon is from their (1972) book *Human Problem Solving* (Englewood-Cliffs, NJ: Prentice-Hall).

The quotation 'new paradigms are needed' is from LeCun, Y., Bengio, Y., and Hinton, G. E. (2015), 'Deep Learning', *Nature,* 521: 436–44.

Minsky, M. L. (1956), 'Steps Toward Artificial Intelligence'. First published as an MIT technical report: *Heuristic Aspects of the Artificial Intelligence Problem,* and widely reprinted later.

Laird, J. E., Newell, A., and Rosenbloom, P. (1987), 'Soar: An Architecture for General Intelligence', *Artificial Intelligence,* 33: 1–64.

Chapter 3: Language, creativity, emotion

MasM, chaps. 7.ii, 9.x–xi, 13.iv, 7.i.d–f.

Baker, S. (2012), *Final Jeopardy: The Story of WATSON, the Computer That Will Transform Our World* (Boston: Mariner Books).

A readable, though uncritical, account of an interesting Big Data system.

Graves, A., Mohamed, A.-R., and Hinton, G. E. (2013), 'Speech Recognition with Deep Recurrent Neural Networks', *Proc. Int. Conf. on Acoustics, Speech, and Signal Processing,* 6645–49.

Collobert, R. et al. (six authors) (2011), 'Natural Language Processing (Almost) from Scratch', *Journal of Machine Learning Research,* 12: 2493–537.

The quotation describing syntax as superficial and redundant is from Wilks, Y. A., ed. (2005), *Language, Cohesion and Form: Margaret Masterman (1910–1986)* (Cambridge: Cambridge University Press), p. 266.

Bartlett, J., Reffin, J., Rumball, N., and Williamson, S. (2014), *Anti-Social Media* (London: DEMOS).

Boden, M. A. (2004), *The Creative Mind: Myths and Mechanisms*, 2nd edn. (London: Routledge).

Boden, M. A. (2010), *Creativity and Art: Three Roads to Surprise* (Oxford: Oxford University Press). This collection of twelve papers is largely about computer art.

Simon, H. A. (1967), 'Motivational and Emotional Controls of Cognition', *Psychological Review*, 74: 39–79.

Sloman, A. (2001), 'Beyond Shallow Models of Emotion', *Cognitive Processing: International Quarterly of Cognitive Science*, 2: 177–98.

Wright, I. P., and Sloman, A. (1997), *MINDER: An Implementation of a Protoemotional Architecture*, available at </ftp:/ftp.cs.bham.ac.uk/pub/tech-reports/1997/CSRP-97-01.ps.gz>.

Chapter 4: Artificial neural networks

MasM chaps. 12, 14.

Rumelhart, D. E., and J. L. McClelland, eds. (1986), *Parallel Distributed Processing: Explorations in the Microstructure of Cognition*, vol. 1: *Foundations* (Cambridge, MA: MIT Press).

The whole volume is relevant; but the past-tense learner (written by Rumelhart and McClelland) is described on pp. 216–71.

Clark, A. (2016), *Surfing Uncertainty: Prediction, Action, and the Embodied Mind* (Oxford: Oxford University Press). A review of Bayesian approaches in cognitive science.

See also the paper by LeCun et al., and the two references from Demis Hassabis' team, cited under Chapter 2, earlier.

The two quotations about the network scandal are from Minsky, M. L., and Papert, S. A. (1988), *Perceptrons: An Introduction to Computational Geometry*, 2nd edn. (Cambridge, MA: MIT Press), viii–xv and 247–80.

Philippides, A., Husbands, P., Smith, T., and O'Shea, M. (2005), 'Flexible Couplings Diffusing Neuromodulators and Adaptive Robotics', *Artificial Life*, 11: 139–60.

A description of GasNets.

Cooper, R., Schwartz, M., Yule, P., and Shallice, T. (2005), 'The Simulation of Action Disorganization in Complex Activities of Daily Living', *Cognitive Neuropsychology*, 22: 959–1004.

Describes a computer model of Shallice's hybrid theory of action.

Dayan, P., and Abbott, L. F. (2001), *Theoretical Neuroscience: Computational and Mathematical Modelling of Neural Systems* (Cambridge, MA: MIT Press).

This volume does not discuss technological AI, but shows how ideas from AI are influencing the study of the brain.

Chapter 5: Robots and artificial life

MasM chaps 4.v–viii and 15.

Beer, R. DS. (1990), *Intelligence as Adaptive Behavior: An Experiment in Computational Neuroethology* (Boston: Academic Press).

Webb, B. (1996), 'A Cricket Robot', *Scientific American*, 275(6): 94–9.

Brooks, R. A. (1991), 'Intelligence without Representation', *Artificial Intelligence*, 47: 139–59.

The seminal paper of situated robotics.

Kirsh, D. (1991), 'Today the Earwig, Tomorrow Man?', *Artificial Intelligence*, 47: 161–84.

A sceptical response to situated robotics.

Harvey, I., Husbands, P., and Cliff, D. (1994), 'Seeing the Light: Artificial Evolution, Real Vision', *From Animals to Animats 3* (Cambridge, MA: MIT Press), 392–401.

Describes the evolution of an orientation detector in a robot.

Bird, J., and Layzell, P. (2002), 'The Evolved Radio and its Implications for Modelling the Evolution of Novel Sensors', *Proceedings of Congress on Evolutionary Computation*, CEC-2002, 1836–41.

Turk, G. (1991), 'Generating Textures on Arbitrary Surfaces Using Reaction-Diffusion', *Computer Graphics*, 25: 289–98.

Goodwin, B. C. (1994), *How the Leopard Changed Its Spots: The Evolution of Complexity* (Princeton University Press).

Langton, C. G. (1989), 'Artificial Life', in C. G. Langton (ed.), *Artificial Life* (Redwood City: Addison-Wesley), 1–47. Revd. version in

M. A. Boden, ed. (1996), *The Philosophy of Artificial Life* (Oxford: Oxford University Press), 39–94.
The paper that defined 'artificial life'.

Chapter 6: But is it intelligence, really?

MasM chaps. 7.i.g, 16.

Turing, A. M. (1950), 'Computing Machinery and Intelligence', *Mind*, 59: 433–60.

The quotations regarding 'the hard problem' are from Chalmers, D. J. (1995), 'Facing up to the Problem of Consciousness', *Journal of Consciousness Studies*, 2: 200–19.

The quotation by J. A. Fodor is from his (1992), 'The Big Idea: Can There Be a Science of Mind?', *Times Literary Supplement*, 3 July: 5–7.

Franklin, S. (2007), 'A Foundational Architecture for Artificial General Intelligence', in B. Goertzel and P. Wang (eds.), *Advances in Artificial General Intelligence: Concepts, Architectures, and Algorithms* (Amsterdam: IOS Press), 36–54.

Dennett, D. C. (1991), *Consciousness Explained* (London: Allen Lane).

Sloman, A., and Chrisley, R. L. (2003), 'Virtual Machines and Consciousness', in O. Holland (ed.), *Machine Consciousness* (Exeter Imprint Academic), *Journal of Consciousness Studies*, special issue, 10(4): 133–72.

Putnam, H. (1960), 'Minds and Machines', in S. Hook (ed.), *Dimensions of Mind: A Symposium* (New York: New York University Press), 148–79.

The quotation about Physical Symbol Systems is from Newell, A., and Simon, H. A. (1972), *Human Problem Solving* (Englewood-Cliffs, NJ: Prentice-Hall).

Gallagher, S. (2014), 'Phenomenology and Embodied Cognition', in L. Shapiro (ed.), *The Routledge Handbook of Embodied Cognition* (London: Routledge), 9–18.

Dennett, D. C. (1984), *Elbow Room: The Varieties of Free Will Worth Wanting* (Cambridge, MA: MIT Press).

Millikan, R. G. (1984), *Language, Thought, and Other Biological Categories: New Foundations for Realism* (Cambridge, MA: MIT Press).

An evolutionary theory of intentionality.

Chapter 7: The Singularity

Kurzweil, R. (2005), *The Singularity is Near: When Humans Transcend Biology* (London: Penguin).

Kurzweil, R. (2008), *The Age of Spiritual Machines: When Computers Exceed Human Intelligence* (London: Penguin).

Bostrom, N. (2005), 'A History of Transhumanist Thought', *Journal of Evolution and Technology,* 14(1): 1–25.

Shanahan, M. (2015), *The Technological Singularity* (Cambridge, MA: MIT Press).

Ford, M. (2015), *The Rise of the Robots: Technology and the Threat of Mass Unemployment* (London: Oneworld Publications).

Chace, C. (2018), *Artificial Intelligence and the Two Singularities* (London: Chapman and Hall/CRC Press).

Bostrom, N. (2014), *Superintelligence: Paths, Dangers, Strategies* (Oxford: Oxford University Press).

Wallach, W. (2015), *A Dangerous Master: How to Keep Technology from Slipping Beyond Our Control* (Oxford: Oxford University Press).

Brynjolfsson, E., and McAfee, A. (2014), *The Second Machine Age: Work, Progress, and Prosperity in a Time of Brilliant Technologies* (New York: W. W. Norton).

Wilks, Y. A., ed. (2010), *Close Engagements with Artificial Companions: Key Social, Psychological, Ethical, and Design Issues* (Amsterdam: John Benjamins).

Boden, M. A. et al. (fourteen authors) (2011), 'Principles of Robotics: Regulating Robots in the Real World', available on EPSRC website: <www.epsrc.ac.uk/research/ourportfolio/themes/>.

Further reading

Boden, M. A. (2006), *Mind as Machine: A History of Cognitive Science*, 2 vols. (Oxford: Oxford University Press).

With the exception of deep learning and the Singularity, every topic mentioned in this *Very Short Introduction* is discussed at greater length in *Mind as Machine.*

Russell, S., and Norvig, P. (2013), *Artificial Intelligence: A Modern Approach,* 3rd edn. (London: Pearson).

This is the leading textbook on AI.

Frankish, K., and Ramsey, W., eds. (2014), *Cambridge Handbook of Artificial Intelligence* (Cambridge: Cambridge University Press).

Describes the various areas of AI, less technically than Russell and Norvig (2013).

Whitby, B. (1996), *Reflections on Artificial Intelligence: The Social, Legal, and Moral Dimensions* (Oxford: Intellect Books).

A discussion of aspects of AI that are too often ignored.

Husbands, P., Holland, O., and Wheeler, M. W., eds. (2008), *The Mechanical Mind in History* (Cambridge, MA: MIT Press).

The fourteen chapters (and five interviews with AI/A-Life pioneers) describe early work in AI and cybernetics.

Clark, A. J. (1989), *Microcognition: Philosophy, Cognitive Science, and Parallel Distributed Processing* (Cambridge, MA: MIT Press).

An account of the differences between symbolic AI and neural networks. Today's neural networks are much more complex than those discussed here, but the main points of comparison still stand.

Minsky, M. L. (2006), *The Emotion Machine: Commonsense Thinking, Artificial Intelligence, and the Future of the Human Mind* (New York: Simon & Schuster).
This book, by one of the founders of AI, uses AI ideas to illuminate the nature of everyday thought and experience.

Hansell, G. R., and Grassie, W., eds. (2011), *H +/–: Transhumanism and Its Critics* (Philadelphia: Metanexus).
Statements and critiques of the transhumanist philosophy supported, and the transhumanist future predicted, by some AI visionaries.

Dreyfus, H. L. (1992), *What Computers Still Can't Do: A Critique of Artificial Reason*, 2nd edn. (New York: Harper and Row).
The classic attack, based in Heideggerian philosophy, of the very idea of AI. (Know your enemies!)

Index

A

B

C

D

E

F

G

H

I

J

K

L

M

N

O

P

Q

R

S

T

U

V

W

Y

ONLINE CATALOGUE

A Very Short Introduction

Our online catalogue is designed to make it easy to find your ideal Very Short Introduction. View the entire collection by subject area, watch author videos, read sample chapters, and download reading guides.

http://fds.oup.com/www.oup.co.uk/general/vsi/index.html